Diagrammatical Performances

Diagrammatical Performances traces some of the traditions and contemporary developments in the uses of diagrams – actual and metaphorical – in the making of performance that nurtures a relationship between humans and more-than-humans.

Addressing students, researchers, and teachers of radical theatre, eco-activists, makers of eco-theatre and site-specific performance, the book provides a radical description of the diagram, a history of its uses in theatre and performance, and a toolkit for its applications in making radical eco-theatre that swerves anthropocentric concerns for more direct engagements with the more-than-human. 'Diagrammatical Performances' draws on a global diversity of practices from those of Atis Rezistans in Haiti to the choreographies of Henry Daniel and Zab Maboungou via the embodied performance art of Ana Mendieta, Miranda Whall, and Louise Ann Wilson. It charts lineages of use through Symbolist Theatre, modernist theatres like Richard Foreman's Ontological-Hysteric Theater, while drawing on ideas from key theorists such as Astrida Neimanis, Baz Kershaw, and Donna Haraway. The book also draws upon the authors' own diagrammatical performance-making in their 'Crab & Bee' partnership – in performances such as 'Around The Hills', 'Leadworking' and 'Scoriton Mysteries' – to present accessible techniques, tactics, exercises and approaches for other performance makers to draw upon.

This book will be of great interest to students and scholars of performance, fine art, and creative writing studies, eco-activists, and media commentators on green matters.

Helen Billinghurst is an independent researcher and interdisciplinary artist.

Phil Smith is a performance maker, founder member of Wrights & Sites, and Honorary Associate Researcher at the University of Plymouth.

Routledge Advances in Theatre & Performance Studies

This series is our home for cutting-edge, upper-level scholarly studies and edited collections. Considering theatre and performance alongside topics such as religion, politics, gender, race, ecology, and the avant-garde, titles are characterized by dynamic interventions into established subjects and innovative studies on emerging topics.

Echoes of Violence on the Early Modern English Stage and Beyond
Edited by Samantha Dressel and Matthew Carter

Performing Waste
To Re-member Pasts and Fabulate Futures
Edited by Dorota Sajewska and Małgorzata Sugiera

Artificial Savages in a Transcultural Landscape
Performance Art, Political Intention, and the Aesthetic Projects of Guillermo Gómez-Peña, Coco Fusco, Violeta Luna, Roberto Sifuentes & La Pocha Nostra
William Stark

Performing Paradise
The Living Theatre in France (1968–2018)
Emeline Jouve

Seven Pillars Acting
A Comprehensive Technique for the Modern Actor
Sonya Cooke

Diagrammatical Performances
Making Theatre with More-Than-Human Others
Helen Billinghurst and Phil Smith

For more information about this series, please visit: www.routledge.com/Routledge-Advances-in-Theatre–Performance-Studies/book-series/RATPS

Diagrammatical Performances

Making Theatre with More-Than-Human Others

Helen Billinghurst and Phil Smith

Routledge
Taylor & Francis Group
LONDON AND NEW YORK

First published 2026
by Routledge
4 Park Square, Milton Park, Abingdon, Oxon OX14 4RN

and by Routledge
605 Third Avenue, New York, NY 10158

Routledge is an imprint of the Taylor & Francis Group, an informa business

British Library Cataloguing-in-Publication Data
A catalogue record for this book is available from the British Library

ISBN: 978-1-032-99911-1 (hbk)
ISBN: 978-1-032-99913-5 (pbk)
ISBN: 978-1-003-60667-3 (ebk)

DOI: 10.4324/9781003606673

Typeset in Sabon
by KnowledgeWorks Global Ltd.

Contents

About the authors

Helen Billinghurst is an independent researcher, author, and interdisciplinary artist. She was awarded her PhD *Ways of making: creating artworks in the studio in response to experiential walking* (2018) at the University of Plymouth and worked as a Lecturer in Fine Art and Painting and Drawing at Arts University Plymouth (2020–2024). Helen has exhibited artworks and presented her research across the UK and internationally. Walking and embodiment are key to Helen's process. Her research interests include place-story, ecology, and folk-culture; she creates multi-sensory re-tellings of old stories, often through an ecofeminist lens. She has written for *Green Letters* (2023), *Performance Research* (2023), and *World Futures* (2020). She was co-editor of *Walking Bodies* (2020), and her poetry and fiction have been published by Triarchy Press, Peakrill Press, and Temporal Boundary Press. Helen has worked in collaboration with Phil Smith as Crab & Bee since 2019; together they explore and reveal the secrets of everyday spaces through artworks, publications, readings, scryings, and performances.

Phil Smith is an Honorary Associate Researcher at the University of Plymouth (previously Associate Professor [Reader]). He was awarded his PhD there in 2013. Before his time in academia, he worked primarily as a freelance playwright and lyricist for numerous small-scale touring companies in the UK, as well as writing for theatres such as TAMS in Munich, St Petersburg State Comedy Theatre and Opera North. In 1980, he co-founded TNT Theatre (now based in Munich) with Paul Stebbings and has been its company dramaturg ever since. In 1997, Phil co-founded Wrights & Sites (site-specific performance and walking arts) with Cathy Turner, Stephen Hodge, and Simon Persighetti. He is the author of 'Making Site-Specific Theatre and Performance' (Bloomsbury), 'TNT: The New Theatre' with Paul Stebbings (Triarchy Press), 'Albion's Eco-Eerie: TV and Movies of the Haunted Generations' (Temporal Boundary Press), and 'Crab & Bee's Matter of Britain' with Helen Billinghurst (Peakrill Press). With Helen Billinghurst he is one half of Crab & Bee, creating diagrammatical performances, exhibitions, installations, poetry, and performative walks.

Introduction

What the book is about

Diagrammatical performance is a practice developed by Crab & Bee (Helen Billinghurst and Phil Smith) in response to the twin crises of climate catastrophe and the hyper-spectacularization of culture. We come from different disciplines; Helen from walking arts, painting, and gallery installation, Phil from walking as a site-specific performance. We did not set out to address these twin crises. However, since we began to work together in 2018, creating a variety of artworks in response to place, these issues have repeatedly stood out for us during our daily walking and art-making practice. Our subsequent diagrammatical performances, and their emergence as a distinctive part of a practice with its own techniques and principles, are our contribution to a dialogue with those crises.

In this book, we will draw on our own experiences and share our own techniques and the theories we draw upon, as well as describe numerous precedents and examples of similarly diagrammatical performance-making from different artists and cultures, now and in the recent and distant past. Our intention is to pass these on as resources for others to draw upon, either in part or in totality, and for others – human and more-than-human – to make of them something for themselves.

Ways to read the book

Multiplicity, assemblage, and heterogeneity are characteristics of the practices about to be described. Therefore, this book is not written to be read like a novel, starting at the first word and ending at the last. Instead, it is organized more like a patchwork, to be read in sections and in any order, according to what you, the reader, need to access – whether that be individual techniques, preparatory activities, overall practice, parallels and precedents, or theory. So, for example, you might want to forego the opening chapter on the theory of diagrams, get straight into the practicalities of diagrammatical performances in Chapter 2, and then come back to the theory. On the other hand, if it meets your needs, follow the printed order.

DOI: 10.4324/9781003606673-1

Here is a very brief summary of what you will find in each of the chapters:

Chapter 1/What are diagrammatics?

This is a theoretical account of what diagrams are in general, and of the particular kinds of diagrams being discussed in this book in relation to performance making.

Chapter 2/Diagrammatics in performance

This chapter gives numerous and varying examples – from different times and cultures – of how performance making has integrated diagrams. It describes the roles diagrams have taken and what uses they have been put to.

Chapter 3/Why is diagrammatical performance relevant now?

Here we describe the contemporary circumstances – from climate crisis to cultural tendencies of transhumanism and destructive futurism – the qualities of diagrammatics are particularly suited to address. There are examples of performances that use some kind of diagramming to attend to these circumstances. The chapter concludes with a discussion of problems in immersive and site-specific performance that diagrammatical performance needs to avoid or resolve.

Chapter 4/Where did our diagrammatical performances come from?

This chapter describes the basis for Crab & Bee's own diagrammatical performance making, rooted in a walking arts practice and an emphasis on "being there" with more-than-human allies. It describes a variety of other arts practices – painting, drawing, and mapping, for example – that can support this performance making, including attention to webs of stories related to places. It describes how the practice of "legend tripping" can be useful.

Chapter 5/Our vision for performing space

This is a detailed account of the critical resources on which you can (and Crab & Bee do) draw: the idea of a positive Planet B, haecceities as alternatives to self-expression, fictioning, changing the ecological cast of characters, hyperobjects and vibrant matter, gravity as an attraction of all things to all other things, salvagepunk, theatrisexuality, and sexy theory.

Chapter 6/Four case studies

After a general account of the creative approach of Crab & Bee, there are four case studies of our recent diagrammatical performances, with critical commentaries: *Wookey Hole* (2024), *Leadworking* (2023), *Scoriton Mysteries* (2025), and *A Séance For Stonehouse* (2023).

Chapter 7/Mini-handbook of techniques

This is a toolkit of actions and approaches that you can use in making your own diagrammatical performances. The sections include: getting started, "being there" and embodiment, researching places, using unrespectable sources, place stories, using a diagram (including drawing, modelling, visualizing), devising with more-than-humans others, the carrier-bag principle, and diagrammatical dramaturgy.

Chapter 8/Conclusion

To wrap up the book, this section looks forward to what kinds of diagrammatical performances might be to come and from what impulses and ideas they might emerge.

1 What are diagrammatics?

What is a diagram?

A diagram is an abstraction and an instruction. It characterizes a situation, a mechanism, or a terrain, but without directly representing its appearance. Occasionally, the thing diagrammatized may be recognizable; where it may seem to be so represented, this is incidental to the work of the diagram.

A diagram abstracts the dynamics or meanings from an object, situation, or assemblage. It strips away those appearances or mechanics that are incidental to the particular dynamics or meanings that the diagram is homing in on. A diagram is unlikely to function as a comprehensive inventory of its subject's various facets. Instead, it will simplify, discriminate, and lift what it takes out of a multiplicitous context in order to isolate motions and meanings from the hyper-complexity of matter and the overwhelming surge of motion in general; diagrams are "reductive renderings ... to concretize process" (Bender and Marrinan, 2010, p. 7). So, for example, a diagram of a water molecule may not include either the appearance of water to the human eye or its operations at the sub-atomic level.

There are certain functionalist diagrams that we can exclude from our diagrammatics (our theoretical resource of diagramming); for example, the diagrams of electrical circuits used to help electricians with wiring are likely to be of little or no use to us. However, what we might find useful is a diagram of how energy moves through a system. For, as well as its reductive qualities, a diagram will often expose and display the connective qualities and the connected allies of what it diagrammatizes (both of which may be of relevance to performers).

Diagrams are not simply made of abstractions. Diagrams are also material things in themselves; a "diagram is not a representation of something else; it is the thing itself" (Netz, 1999, p. 22). They have a dynamism that is distinct from the dynamism that they represent; they are "things to work with" (Bender and Marrinan, 2010, p. 10). Due to their materiality (which is perhaps most evident when they are hand-drawn), they can be *directly* deployed in a performance to give shape to a performance space, to a scenography, or to embodied gestures. However, it is more often their

DOI: 10.4324/9781003606673-2

blend of abstraction and materiality – constituting a dynamism within the diagram itself – that is of particular interest to us, given that it energizes signs and symbols with material presence, potentially manifesting them as agents in the performance space, at the same time as connecting objects in the performance space to vast webs of meaning beyond. While diagrams can operate in multiple dimensions (including temporal ones), they "flatten" complex objects, aiding them in spreading their webs, meshes, or networks far and wide. Despite their abstract quality, diagrams can assist us in working ecologically, given their tendency to connect in complex ways and to discriminate in favour of relations over objects (hence the importance, described below, of objects that can bring the materiality of an ecological hyperobject to the diagram). A diagram joins discrete objects to threads of communication; it associates a thing here with a hyperobject that extends over there and beyond our reach. It contextualizes any utterance to a tradition of utterances.

Non-functional diagrams

In his 2016 essay *On The Diagram (and a Practice of Diagrammatics)* Simon O'Sullivan (pp.13–25 in Schneider & Yasar), academic and performance maker with Plastique Fantastique, identifies a number of non-functionalist diagrammatical practices – from the sigils of occult artist Austin Osman Spare to philosopher-psychologist Jacques Lacan's *graphe du désir* (diagram of desire) to his own use of diagrams such as his "diagram of activation of device" (see *figure 1.1*) – not only in order to define a certain kind of diagram, but also to propose the affordance of such diagrams for constituting "protocols for a possible practice (of diagrammatics)" (p. 13). O'Sullivan energizes this proposal by drawing on a further dynamics, which is the tension between the diagram at its most abstract – in which the process of abstraction (in fictions, forms and signs) is able to float free of what it abstracts by parasitically robbing its subject of its meaning in order to "communicate without meaning" (p. 14) – and the diagram as a "creaturely" or animal entity that "foregrounds the potentialities of the body and world" (p. 16).

This might seem hard to imagine in conceptual terms, but in the context of a performance, it becomes far clearer. For example, if a ceremonial magician (in a variety of cultures) marks out a circle, they are often drawing on all sorts of abstract ideas of eternity, the Oroboros, self-feeding energy and so on, at the same time as physically demarcating a space of protection and intensity with, say, salt or ashes, in order to contain or exclude certain forces or more-than-human personalities. Similarly, the mudras (hand gestures) used in Kathakali dance theatre are not simply in a one-to-one (sign-to-signified) relationship, but are, more significantly, parts of a physicalized and highly diagrammatized "creaturely" invocation of an organization of the cosmos that is, itself, both abstract and "animal," "where the principle of the 'unmanifest' and the 'varied manifestations' was embodied" (Vatsayan, 1980, p. 9).

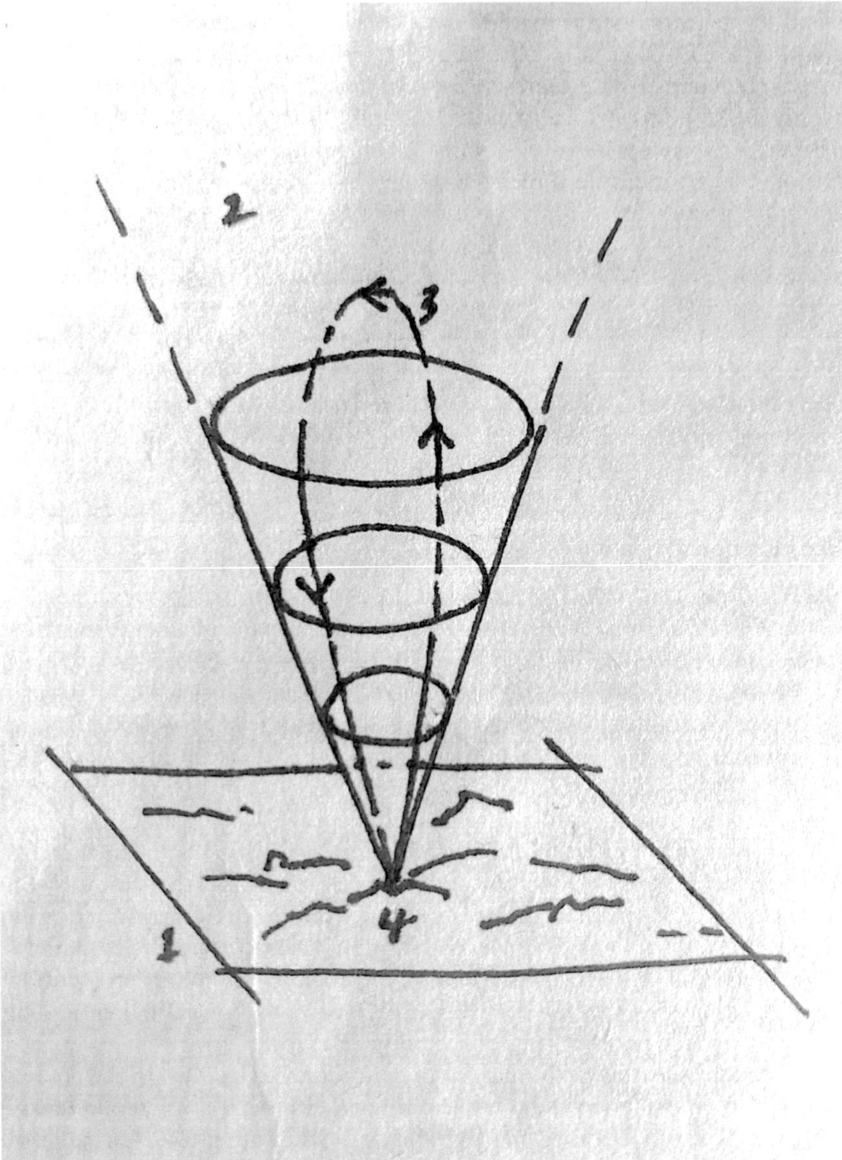

Figure 1.1 Simon O'Sullivan's "Diagram of activation of device" from "From Magic and Myth-Work to Care and Repair," 2024, p. 52.

This reciprocal embodiment in diagrams – of the "unmanifest" invisibles of the cosmos and the material body of the performer – is navigated in post-structuralist thought (often via the philosophy of Baruch Spinoza) as a body "constituted by relations of slowness and speed that themselves define its capacity to affect and be affected" (O'Sullivan, 2018, p. 173). While there is an obvious relation here to everyday human ethics (how a body affects other bodies), what the diagram does is to level out the human with the more-than-human, allowing the performer to imagine and perform becoming (and attending to) the more-than-human. The radical consequences of this for a relation to the ecological "meshwork" (Ingold, 2008, p. 3), the subject of ecology, should be self-evident.

Protocols for a practice

Thus, diagrammatics as a practice (or "protocols" for a practice) tend to both the performative and to the transformative; for they operate at the edge of what is material and what is not, passing back and forwards in a way that is likely to affect the constitution (and metabolism) of whatever passes from one to the other. In this respect the diagram is generative; a quality highlighted by curators Karin Schneider and Begum Yasar when they contextualized their 2016 "Situational Diagram" exhibition at the Dominique Lévy gallery in New York as responding to a crisis in the justification for art making in a "world of ever-increasing inequality, of ceaseless injustice and conflict". They characterized their curation (and supporting publication) as a means to a renegotiation of "our relationship to abstraction and to images in the wider sense [as] a political act" through the "reinvention of our subjectivities" (Schneider and Yasar, 2016, p. 9).

In a performance using diagrams, then, a performer should expect a reciprocal use of themselves by the diagram. The artist's subjectivity (or creativity) may not be the primary agent or material in play, but it will be one of many agents and materials all subject to each other. On the diagram the artist is not necessarily any more in control than any other node. While a diagram can assume a multi-dimensional and model-like shape, it tends to be horizontal and so can flatten hierarchies. Even where it is multi-dimensional, a diagram is always likely to be subject to meta-modelization (a modelling of models) in which privileged aspects of any one model are likely to lose their priority across the diagrammatized field. Equally, using diagrams affects the artist not only in challenging their status in relation to their materials (their more-than-human collaborators) but also in flattening the different conceptualizations of their subjectivity and its parts, so that those parts usually privileged in dominant popular discourses – will, drive, self-consciousness – may be lowered into more equitable relations with usually less valued aspects such as the unconscious and intuition.

Psychologist Felix Guattari (often in collaboration with philosopher Gilles Deleuze) is one of a number of writers in the fields of neo-vitalism and post-structuralism (although they mostly defy categorization) we have repeatedly drawn on. He uses a four-sided diagram to describe "meta-modelization". Along the axes of "expressions" and "content" on one side, and of "the possible" and "the real" on the other, Guattari places symbols for four fields:

the constitution of any entity, the ambiguity of the chaos and cosmos ("chaosmos") from which any entity emerges, the autopoietic capacity of that entity (its ability to produce its parts from itself), and the complexity of the relations that sustain that capacity (Guattari, 1995, p. 60). By using such a diagram, as performance makers we can avoid ignoring the agency of all the "apparently" inanimate entities around us, and instead make queer and sensual alliances with them, while plunging our own agency and subjectivity (hopefully with some kind of containment) into the chaos part of the cosmos.

If this nesting of categories/fictions is "doing your head in," then it is working. For part of the point of these nestings, one material/abstraction within another, is to allow an undoing and a re-doing of our usual relations with matter and motion, and to deterritorialize and then reterritorialize our boundaries and connections (Deleuze and Guattari, 1988, p. 321) through a brief encounter with the "chaosmos". This process seizes on the quality of the diagram – always partly enigmatic due to its simplifications and abstractions – to require our responding by re-ordering or supplementing its information: its "disunited field of presentation – ruptured by shifts in scale, focus, or resolution – provokes seriated processes demanding an active correlation of information" (Beneder and Marrinan, 2010, p. 8). The diagram demands we reorganize its information, meanwhile insinuating itself into a reorganization of us.

While this may have a profound impact on the life of the artists involved, in diagrammatical performance there is no diagrammatical requirement for the kinds of "individual life" that characterize dramatic theatre. In diagrammatical art and performance, stories can be recounted in the first person as if in the third. Subjectivities are present as distant points on the web. This releases a performance maker from the kinds of emotional investment essential to drama. Where there is none invested or at stake, whether fictioned or actual, cultural capital is less likely to be accumulated. Extractive value is displaced, not only from any residual drama but also from the diagram and the performance to other points on the web: geological, geographical, and mythical. There is no mimesis; nothing needs to be imitated when materials can be present. In a diagrammatical performance, there is no need for "acting" in the sense of embodying a character. "Characters" are already present across the diagram in the form of more-than-human agencies. Everything can be presented rather than represented. If there is any embodying of character, it is more in the manner of a ceremonial invocation; without a need for imitation, a magician (ancient or contemporary) would use abstract gestures (an embodied diagram) and non-representational written signs, sigils and patterns (a creaturely diagram) with "sympathetic" substances to invoke some absent entity (god, spirit, ancestor) to present itself. So, for example, if a Hindu idol is formed from mud from the roadside, there is no imitation of a god, but for a while the mud idol *is* the god – the god is present rather than represented – until the idol is broken up and the mud is returned to the roadside and the god has gone. Mud remains.

The double qualities of the diagram

In this way, a diagrammatical performance always loops back in time to its origins in a "being there" with more-than-human others. It is the nature of both the connectivity of the abstract diagram and its paradoxically creaturely qualities that the more-than-human is always as present as any human subject. Once a diagrammatical performance ends (the idol is broken up), all focus and connection should return to the places and the agencies where it began; so the looping continues. This kind of performance is thus partly beyond art, and part instead of a situationist "art of living": "'The realisation of art – poetry in the situationist sense – means that one cannot realize oneself in a "work," but rather realizes oneself period'. After art comes the art of living" (Black, 1995, p. 86).

Despite the connectivity of the diagram and its capacity to be part of a transformation of everyday life, a diagrammatical performance is unlikely to establish any kind of wide dispersal as a practice. For it is at the mercy of a wider emergence (or lack thereof) of an "art of living" into general social practice, which may, even then, adopt or ignore it. In terms of cultural discourse, the diagram, by its dispersal of values to an elusive materialism of "swerves" (Nail, 2024, pp. 38–41), will always be unattractive to a commerce in transactional products. In that sense it is an "avant-garde" practice out of joint with its time (the era of the disappearance of "avant-gardes" [Ley, 1991, pp. 348–352]). Instead, it is closer to the practice of certain "minor literatures" and experimental practices that ignore instrumentalism and social relevance in favour of intensity, and resistance to the efficiencies of the mainstream and the auto-generation of perverse products.

While Deleuze and Guattari described many such minoritarian processes – "All becoming is minoritarian" (Deleuze and Guattari, 1988, p. 117) – they rejected any need to visually represent diagrams *as* them. Or as *par*t of them. Instead, they "stresse[d] the morpho-genetic capability of matter to generate form on its own" over the human drive through representation "to impose [form] onto it from outside" (DeLanda, 2000, p. 33). In other words, although we may use diagrams to make connections to, and generate fields on which to meet, with our more-than-human collaborators, we have no right to expect those collaborators to conform to our diagrams. There is, however, no reason why the mutant products of their autopoiesis – in response to or rejection of our diagrammatical thinking – cannot make a physical appearance in our performances.

References

Bender, J. and Marrinan, M. (2010) *The culture of diagram*. Stanford, CA: Stanford University Press.

Black, B. (1995) *Beneath the underground*. Portland: Feral House.

DeLanda, M. (2000) 'Deleuze, diagrams, and the genesis of form', *Amerikastudien/American Studies*, 45(1), pp. 30–34.

Deleuze, G. and Guattari, F. (1988) *A thousand plateaus*. Translated by Brian Massumi. London: Continuum.
Guattari, F. (1995) *Chaosmosis*. Translated by Paul Bains and Julian Pefanis. Bloomington & Indianapolis: Indiana University Press.
Ingold, T. (2008) 'Bringing Things To Life: Creative Entanglements In A World Of Materials'. *Department of Anthropology, Aberdeen University*. Available online at: http://eprints.ncrm.ac.uk/1306/. Accessed: 11.1.2026.
Ley, G. (1991) 'Sacred "Idiocy": The avant garde as alternative establishment', *New Theatre Quarterly*, 7(28), pp. 348–353.
Nail, T. (2024) *Matter and motion*. Edinburgh: University of Edinburgh Press.
Netz, R. (1999) *The shaping of deduction in Greek mathematics: A study in cognitive history*. Cambridge: Cambridge University Press.
O'Sullivan, S. (2016) 'On the diagram (and a practice of diagrammatics)', in Karin Schneider and Begum Yasar (eds.), *Situational diagram*. New York: Dominique Lévy.
O'Sullivan, S. (2018) 'Memories of a Deleuzian: To think is always to follow the Witches' flight', in Henry Somers-Hall, Jeffrey A. Bell and James Williams (eds.), *A thousand plateaus and philosophy*. Edinburgh: Edinburgh University Press.
Schneider, K. and Yasar, B. (2016) *Situational diagram*. New York: Dominique Lévy.
Vatsayan, K. (1980) *Traditional Indian theatre: Multiple streams*. New Delhi: National Book Trust, India.

2 Diagrammatics in performance

Precedents, parallels, and recent applications

Ancient and traditional performances

Diagrams are far from a constant presence in performance, yet their use has been common, even fundamental – "pivotal to theatre" (Mirčev, 2024, p. 1) – at certain times and in particular cultures. They have appeared in ancient rituals that are only now partially recoverable by archaeological excavation or in textual revenants, such as the diagrams used in the serpent-reverencing rituals of the Ophite sect of second-century (CE) Gnostics (Welburn, 1981, pp. 270–272) who were primarily based in Egypt and Syria. Or as the *zisurrû* (magic circles defining efficacious spaces of performance, of healing for example) of ancient Mesopotamia (Smith, 2011, pp. 198–201). Yet, diagrams have also been a mark of trends in modernist performance – such as in the deployment of "picturization" (Kirby, 1973, p. 11) in the Ontological-Hysteric Theater (founded in 1968) of Richard Foreman – and in contemporary performances.

In some cases, diagrams are still explicitly deployed in longstanding traditional performances, such as the *Gaan* dance of the White Mountain Apache Crown Dancers of the Mescalero Mountain Spirit tradition (Ball, 2000, p. 268). Embodying *Gaan* (mountain spirits from the caves), the four black-masked performers of this dance wear diagrammatical headdresses representative of the fourfold directions of everything (the exact symbols for which the dancers choose themselves, though often these include the "Apache cross" and symbols representing natural forms). A white-masked dancer shows the other four in the dance "the way to live, what's in the forest, how to use it, how to pray" (Tohonnie, 2012, unpaginated). The detailed patterning of the dancers' body paint and "crowns" headwear contrasts with (or, perhaps, balances) the non-choreographed nature of the dance, which the performers intuit from the accompanying song and drumming. While all such conservational performance practices are subject to change (according to their leader, Joe Tohonnie, the Crown Dancers' costumes and body painting have become more detailed in recent years), the intention of these dancers is to preserve and pass on a tradition. Diagrammatics may well assist the longevity of their efforts, preserving a pattern of performance as its ornamental quality changes.

DOI: 10.4324/9781003606673-3

The resilience of such diagram-based performances is hard to chart exactly, but there is some evidence to suggest that these may persist over very long periods. A recent archaeological study at Cloggs Cave on Gurnaikurnai land (in Australia), at the invitation of Elders there, uncovered ancient twinned upper and lower fireplaces and twin sticks treated with animal or human fats. This arrangement (a diagrammatical relation) appears to replicate descriptions of preparations for a Gurnaikurnai healing ritual recorded by a government ethnographer in the nineteenth century. If this similarity is a result of transmission rather than coincidence, then that implies a longevity of up to twelve thousand years, or five hundred generations (David *et al.*, 2024, unpaginated).

If a relation between diagrammatics and attention to a wider ecological web can be deduced from the presence of animal fats in Cloggs Cave, such a sensibility appears even more clearly at the "Rhino Cave" in the Tsodilo Hills (*Ju/hoansi San*), Botswana. Here, recent archaeology suggests a ritual (of a similar or earlier dating to that at Cloggs Cave) that was centred on a fortuitously shaped or enhanced rock which "can be seen as a zoomorphic form, such as a snake or a tortoise head.... [that] 'rears up'"; "the use of a particular rock panel as a prop brings to life a seemingly 'dead' cave wall; it becomes a living entity where the rest of the body is inside the wall" (Coulson, Staurset, and Walker, 2011, pp. 49–50). The presence of flecks of ochre in the cave and evidence of the burning of numerous unused "points" (spear or arrow tips) suggest that there was a striking red colouration to whatever took place there (p. 49). While this archaeology has been subject to excited journalistic hyperbole (Coulson, 2008), there remains a possible continuity of transmission to the numerous snake myths still current among the KhoeSan people living around Tsodilo Hills today (Sullivan and Low, 2014).

Diagrams of resistance

While these performances are subject to the likelihood of accidental variations and organized reformations and restorations, there are also examples of contemporary transmissions of longstanding diagrammatically-based performance that are being self-consciously adapted. One of these is the Vodou based work of the street sculptors of Atis Rezistans at Grand Rue in Port-au-Prince, Haiti. Rooted in ritual service to spirits (*Lwas*), qualities of longevity, resilience, and contemporaneity in performative practices are brought together in an assemblage for attention to the spirits that includes ritual repetition and improvisation.

The patterns of performed Vodou veneration, in general, include short sections that are repeated, temporally and spatially, in a "cyclical ritualizing" (Hebblethwaite, 2012, unpaginated). The street sculptors of Port-au-Prince have incorporated a rhythm of recycling into their practice. In their 2009 Ghetto Biennale, an event founded by Andre Eugène of Atis Rezistans and curator Leah Gordon, numerous *Gédé Lwas* (spirits, or representations of spirits, of the *Gédé* "nation") were gathered in a mechanical-supernatural chorus under the looming figure of Eugène's *Bawon Samdi* (a manifestation of Bawon Samdi [Baron Samedi] who with Gran Brigit [or Mamam Brigitte] leads the family of *Gédé Lwas*). These figures – the work of numerous performative sculptors from the Grand Rue area – are only ever approximately humanoid in form; sometimes not at all. They are expressive of the personalities of the *lwas*, but equally so of the broken and discarded detritus of the streets from which they are made (including parts of motor car chassis, scraps of wood, wiring and fan belts) and in turn of the "make-do tactics, get-by strategies, ad hoc tools, and craftily wrought machines" necessary for survival, let alone flourishing, in the city from which they arise (Braziel, 2016, p. 419). They are frame-like, full of holes (portals to unearthly realms), they are patterned structures onto which to hang things; modest machines ready to interact with a spectator; some are fed at festivals, while the sprung penis of Eugène's *Bawon Samdi* will respond to touch.

These spirits are literally of the matter of the streets; they share in its recuperative economies. Those who are possessed by *lwas* at ceremonies will trade in obscenities (*betiz*) and erotic dance (*goyad*). Resistant to the good taste and control of the official Church, forged as Vodou was in a successful resistance to slavery, it now provides a recuperative economic, erotic, and wise drive to communities deprived of access to local natural resources and forced to live off and transform the trash of wealthier societies. At the same time the figures of the *Gédé* family diagrammatize an unseen but ever-present realm of the dead; they place their makers, co-performers and spectators at a point of connection between two realms, their lineage in "the long, historical, cultural reservoir of the African diasporic religion" (Braziel, 2016, p. 419) interwoven with figures and symbols of colonial Christian Catholicism.

Multi-sensory engagement with the Vodou structured cosmos is explicit within the many complex drum rhythms, and corresponding anti-rhythms called *kase* (meaning "break") that are used to call in the *lwas* (Schwarz, 2017, unpaginated). The patterns of their *vévés*, created by priests and initiates in coffee or flour, set their personalities in cosmogrammatical forms, and conceptually shape their roles in performance. To participate in the Vodou diagram is to experience both order (guardianship, care) and chaos (disordering of identities, public confessions, daunting drives). Like the diagram itself, Vodou is abstract and creaturely. In an exhibition/performance at the ALPHA60 gallery space in Melbourne in 2023 (see *figure 2.1*), Jean-Daniel Lafontant – Vodou priest and guardian of Temple Na-Ri-VéH 777 – first performed a blessing and vivification of the *lwas* sculptures on display, then supervised the making of their sigil-like *vévés* (see *figure 2.2*), which makers and spectators then "destroyed" in ecstatic dancing, leaving an array of vast abstract images "with profound mediumistic qualities" (RISING, 2023, unpaginated). The relation of performer and "content" in Vodou is alarmingly reciprocal; as Jean-Herard Celeur says in the film *The Sculptors of Grand*

Figure 2.1 Installation view of *Ghetto Biennale* by Atis Resistans, RISING (2023). Commissioned by RISING Melbourne. Artists: Katelyn Alexis, Edouard Duval-Carrié, André Eugène, Leah Gordon, Bastian Hagedorn, Henrike Naumahn, Laura Hayman, Michel LaFleur, Herold Pierre Louis, Evel Romain, Claude Saintilus, John Cussans, Roberto N. Peyre. Ghetto Biennale Curators: André Eugène and Leah Gordon. RISING Curator: Grace Herbert. Photos: Ian Laidlaw.

Rue (Leah Gordon, 2008): "I am Gede. It's a Gede that's speaking to you. I've already died."

When performed in a Vodou temple in Haiti, both the choreography of the dances and the architectural/scenographic layout are simultaneously diagrammatic and functional. The central pillar (*potomitan*) of the temple, for example, functions as a conduit for the *lwas* but also as a hinge-point for the rituals, which are themselves cyclical and repetitive, returning again and again to songs that summarize the Vodou cosmos in short verses: "cycles of drummed and danced ritual salutations … In highly formalized and choreographed worship" (Hebblethwaite and Weber, 2016, p. 227). By returning repeatedly to the same "stations" of the temple, including the *potomitan*, the practitioners salute, libate and honour each space, laying patterns of choreography, liturgy and sacred architecture one over another.

Similar patterns and relations are at work in the performances, elaborately diagrammatical costuming and masks, and driving rhythms of the Afrofuturist music collective Fulu Miziki from Kinshasa in the Democratic Republic of Congo. Their name translates from Lingala into English as "music from

Figure 2.2 Installation view of *Ghetto Biennale* by Atis Resistans, RISING (2023). Commissioned by RISING Melbourne. Artists: Katelyn Alexis, Edouard Duval-Carrié, André Eugène, Leah Gordon, Bastian Hagedorn, Henrike Naumahn, Laura Hayman, Michel LaFleur, Herold Pierre Louis, Evel Romain, Claude Saintilus, John Cussans, Roberto N. Peyre. Ghetto Biennale Curators: André Eugène and Leah Gordon. RISING Curator: Grace Herbert. Photos: Ian Laidlaw.

garbage"; they create music and fashion by upcycling found objects and have a concern for the environment at the core of their performing. "We hope our collective can put a spotlight on this pollution problem in Kinshasa … finding solutions is what Afrofuturism is about," according to collective member Skelembele (McCool, 2022, unpaginated). Fulu Miziki takes the same approach as the sculptors of Grand Rue to their musical and performance composition, combining a reclamation of abandoned materials with a welcome to the close influence of the dead. In the words of band member Bokungako Deboule:

> We tried to get some new sounds in our ears…. we looked at things that people threw away and thought we could get some sounds from rubbish…. when we put on costumes we call it "transmutation"; channelling the people who first had the idea of musical instruments long before guitars were invented. Those people who lived before inspire us now, their spirits are incarnated into us and they push us to do this. That's why they call us futurists
>
> (ABC Arts, 2024, unpaginated)

Modernist diagrams

An overtly modernist performance deploying diagrammatics emerged in the USA and Western Europe in the second half of the twentieth century. Richard Foreman's Ontological-Hysteric Theater, formed in 1968, was founded explicitly on ideas of consciousness and time, and theoretically indebted to Gertrude Stein's concept of a "continuous present" based on William James's ideas about "streams of consciousness" as the means to introspective reflection and analysis of what is "sensibly continuous" (Davy, 1978, p. 110). Foreman sought to temporally flatten the psychological terrain of his theatre's performances by removing any sense of narrative progress or dramatic development. He created stage "pictures" akin to diagrams (occasionally with projected texts, like the labels of a diagram) in that they depicted not mimetic representations of the appearances of embodied social interactions, but abstracted diagrams of the machinic processes of connecting and image-constructing within the "continuous present" of such interactions.

For Foreman's production of his self-scripted *Sophia = Wisdom Part 3: The Cliffs* (1972–1973), critic Michael Kirby describes a "picturization" by which performers "present essentially two-dimensional images to the spectator" for which "[T]he props and furniture.... are aligned parallel to the audience" (1973, p. 6), "[E]verything is closed in, arranged, and clearly organized.... The space is arranged and divided purposefully and intellectually" (p. 7), "scenes are presented as sequences of static pictures" (p. 10), and "the picture plane.... emphasized" (p. 11).

In the second act of the play, the character of Karl has one end of a piece of string placed in his mouth; the other end is held by another character reclining on the floor. A ribbon, attached to Karl, is then run through a hole in a small screen placed in front of Karl and then run offstage and out of sight. Performers hold thick stole-like ribbons (one end of which is held by Karl) connected to T-shaped structures. At the ends of the crossbars of the T-shaped structures are boxing gloves. A young female character is seated, posed, in front of Karl, looking away from him. Two performers stand on either side of the T shapes; when they begin to "shadow box," another performer enters and records in chalk the foot movements of one of these "boxers" (pp. 12–13).

The effect is a diagram (occasionally in sedate motion, but mostly still) of social and psychological interactions between characters who do not explicitly interact. Stripped of a representation of the appearance of social interactions, what remains are the trajectories of desires and drives of the "characters" in abstract but overt ("creaturely") forms. Foreman was all the time visible throughout the performances of *Sophia = Wisdom Part 3: The Cliffs*, close to the front of the performance space, operating a tape recorder and determining the pace and rhythms of scenes – as the Crown Dancers' singer/drummer determines their dancing – at times verbally cueing actions, so that the fabrication of the performance was rendered as "bare" and brazen as the psychological drives.

The diagrammatical nature of *Paradise Now* (1968), performed by Judith Malina and Julian Beck's Living Theatre, was explained by Malina. She described how it consisted of "Ten Rungs, leading upwards from The Rite of Guerrilla Theatre.... to the highest vision of the permanent revolution. Each rung consisted of a Rite, a Vision, and a Political Action in which the audience took part.... What looked like chaos was in fact a free-flowing movement toward more and more elevated forms" (Malina, 2012, p. 181). In Living Structures'*Leviathan* (2014–2016), based on Herman Melville's *Moby Dick*, the company's diagrammatical approach to the novel's text effected an abstracting of certain orderly structures, colours, volume, mass, referencing Constructivist aesthetics which were then deployed in interactions with depictions, in sound and visual image, of the vast formlessness of the fiction's white whale (see *figure 2.3)*.

In all the above examples, ancient or modern, there was a literal use of diagrams. At times their performance spaces, or parts of them, would look like a diagram; in *Paradise Now* bodies lay across the performance area in diagrammatic shapes, while the Living Theatre's *Frankenstein* (1965) tipped the performance space through 90 degrees and diagrammatized it as a scaffold grid, across the units of which the acting company formed images, both symmetrical and organic. However, there are other cases where there may not be any immediate visual or material evidence of diagramming, but rather it is the dramaturgy of the performances that is diagram-like.

Figure 2.3 *Leviathan*, Living Structures, 2014. Hackney Downs Studios (UK). Directed by Klaus Kruse. Performers: Dugald Ferguson, Theodora Van Der Beek, Yolande Bramble-Carter, Verity Standen, Jenny Minton, Rosa Iwin Clark, Eleanor Showering. Photo: Ambra Vernuccio.

Examples of this proliferated in late nineteenth and early twentieth century Russian and European Symbolist theatre when Symbolist dramatists sought to manifest the invisible as a reaction away from both the sentimentalized or sensationalized emotionalism of contemporary melodramas and the emergence of a secular realism. For example, in Maurice Maeterlinck's 1891 play *L'Intruse* (*The Intruder*), a stranger, "Death," enters a house where a woman is dying and her family has gathered. This stranger is never seen but is experienced by the residents as sounds and changing moods. The play's stage directions indicate a naturalistically represented room. However, the action of the play renders this apparently material setting porous; the house is effectively helpless against the arrival of "the intruder" who invades the space conceptually and with physical consequences. In Villiers De Lisle Adam's play *Axël* (1890), the central characters explicitly strive for the invisible. After rehearsing and abandoning various forms of esoteric and ceremonial initiation as insufficiently spiritual, they renounce life itself as inadequate for their transcendental requirements: "Vivre? Les serviteurs feront cela pour nous" ("Life? The servants can do that for us"). In both cases, *L'Intruse* and *Axël*, what the audience sees and hears is secondary to the implied diagramming of death and transcendent (spiritual) living.

Characterized by a lack both of dramatic action and of anything material being significantly "at stake," contemporary performance makers staging Symbolist texts were often drawn to something like an overtly diagrammatical staging. In these productions the invisible or esoteric content of the drama (often influenced either by Hermetic philosophy or by mystical Eastern Orthodox Christianity) could be represented or symbolized in patterns of staging, such as the tableaux deployed on a very shallow stage by Vsevolod Meyerhold in his 1906 production of Maeterlinck's *Sister Beatrice*. Similarly patterned staging has been evident in more recent Symbolist-influenced theatre such as Robert Wilson's *Einstein On the Beach* (1976) and Odin Teatret's *The Gospel According To Oxyrhincus* (1985).

The compartments of happenings

In 1952, an untitled evening of performance at the Black Mountain College in North Carolina, USA was to spawn a new genre of performance: "the happening." As a preface to this performance, composer and pioneer of indeterminacy in the arts, John Cage, read out a passage from a Zen Buddhist text: "Art should not be different (from) life but an action within life.... with its accidents and chances" and then began to turn a radio on and off according to time segments in a score. The artist Robert Rauschenberg played records on a hand-wound gramophone, and a pianist played on a piano, the hammers of which had been interfered with. Water was repeatedly poured from one bucket into another, two poets read their work over each other's, and dancers led by the choreographer Merce Cunningham danced through the aisle pursued by a playful dog.

The preparation for this "happening" – which became a model for subsequent performances – was a mixture of precise structuring and a refusal to prepare discursive content, with an insistence on improvisation and ordinariness within the designated compartments of a prearranged temporal and spatial structure. A theme was (sometimes) chosen by the organizers of these "happenings," and a "score" was composed that delineated and apportioned times and locations as instructions for performers during a performance. How the performers filled their allotted time and space, however, was up to them. Elements of a popular mass-production culture of movies, cars, and rock n' roll often manifested, but "happenings" were often attempts to disrupt the logic of that culture. The nature of what individual performers or groups intended to do within their allotted compartments was kept from the other performers; any resulting juxtapositions and their meanings were fortuitous.

Absent from these performances was any conventional story-structure based on tension, conflict, crisis, climax, and resolution. Gone was any structure that informed an audience how to feel or what to understand. In a "happening" the meaning had to emerge or be intuited or fabricated by the spectators as it was performed, and the performance was opened up for circumstance and happenstance to play a role within it.

What unites the subtler diagrammatical dramaturgy of Symbolist performance with the "happenings" and their more literal use of the diagram – in designated compartments of time and space – is that all these performances were almost always expressive of a conscious *weltanschauung*, a world-view or cosmic-view; religious, gnostic, animist, psychological, structuralist, post-structuralist or counter-cultural. In different ways, they addressed a *cosmos* or ecological meshwork, connecting to larger patterns or hyperobjects via the "lines of flight" of their diagrams.

Diagrammatical choreographies

A more recent search for such a relation, to an idea of cosmos through diagramming in performance, is evident in the work (ongoing) of choreographer Henry Daniel. In his performance *Touched* (2008) as part of *The Touched Project* (2007–2010), working with two engineers and a mathematician, Daniel diagrammatized the stage with a symmetrical low-walled arena in which semi-autonomous robotic machines (floor cleaners) were set in motion. Reflecting Daniel's interest in complexity and Chaos Theory, the machines were programmed to repeat predictable patterns which when the machines interacted with each other generated unpredictable moves, to which in turn the human performers of *Touched* responded.

At the same time, there was a "simpler" complementary diagramming of space in *Touched* (see *figure 2.4*). A ramp to one side of the stage provided a platform for a personification of Shiva, the renewer, initiator of dance, divine creator and destroyer, and, beside it, space for Shiva's avatar, at times facing

Figure 2.4 Dancer Jeanette Kotowich and cast with "rumbas" (electronic vacuum cleaners) in Henry Daniel's *Touched*, premiered April 3, 2008. SFU main-stage theatre, Burnaby, Canada. Photo: Henry Daniel.

Shiva, at times turning away. Thanks to this material and explicit dynamic of doubleness beside the complexity of the performers and floor cleaners, the performance can be conceptualized (but not necessarily hierarchized) in terms of three shifting layers (as described in a talk by Henry Daniel at University of Plymouth, UK, 5.2.25): a human layer, a divine layer, and a zone of technological intermediaries. The connections between these layers were animated by reference to the meeting space of the depicted fingers of Adam and God in Michaelangelo's Sistine Chapel Ceiling (Daniel, n.d., unpaginated).

In *Touched*, and in his work more generally, Henry Daniel has swerved controversies around liveness, (including Philip Auslander's assertions in his book *Liveness* [see Auslander, 2008] about the redundancy of liveness in the face of new reproductive/representational technologies), in order to better address presence and time by "creating a connection to the Futurist idea of simultaneity as a non-linear approach to time and space and the practice of juggling multiple domains and personalities with the notion of a fractured self that had a distributed presence in multiple locations" (Daniel, 2022, p. 126). To do this in *Touched*, Daniel created a "non-site" for performance: a three-dimensional and material metaphor (Smithson, 1996, p. 364), across which an idea proposed by choreographer Zab Maboungou – that complex polyphonic rhythms have a "capacity to free the body from the constraints of limited earthly existence" (Daniel, 2022, p. 126) – could be worked out through the various material and conceptual zones in a dance between humans, machines and divinities.

In Zab Maboungou's own work a similar diagrammatical relation to Henry Daniel's three layers and doubling is evident in its relation to the circle. The circle is something which, for Maboungou, is too often used as a "tendency to circumvolution" in dances based on indigenous performances. She detects something else at work within and beside it: "If you look at those ancient, traditional dances.... you'll find that geometry is at the heart of what looks like a circle.... the rhythms themselves call for geometry, lines in space, architecture, points.... You can very well make a square and yet be in a circle.... you need the circle. It's the breath that is circling" (Montréalserai, 2021, unpaginated). So, here the conceptual, or at least the invisible in a spiritual layer or as "breath," works diagrammatically alongside the visible patterns in performance.

Crab & Bee have mostly avoided using live technologies (instead, deploying parts of machines in ruins and the trashed products of mass production systems), locating ourselves in the waste of human production, in a post-industrialism that includes an anticipated digital precarity. Henry Daniel's openness to numerous belief systems (making connections rather than discriminating in terms of conflicting values or distinguishing through contradictions), extends to science and technology. His work with "robotic systems, interactive sound manipulated in an interactive context and sensor-driven motion-capture technology" (Daniel, 2022, p. 126) suggests that live technologies can be drawn into the diagramming of a performance, facilitating multiplicitous connections rather than simply mediatizing the animation of a relation to cosmos. That this relation is transcendental, as it seems to be in Maboungou's and Daniel's cases, may indicate that there is an affordance for the digital in a transcendental diagram.

This raises questions that may engage researchers in diagrammatical performance in the future. The first is to what extent any re-use of diagrams taken from ceremonial magic or "ancient, traditional dances" can or does carry cultural and other meanings across from their original contexts. A second issue – closely related to the first, given how, for example, contemporary Chaos magicians have embraced elements of digital technology (Lachman, 2018) – concerns a relation between the algorithmic or programming qualities of digital technologies and the creaturely abstractions of diagrammatical performance.

These are not necessarily recent concerns, dating back at least as far as the first use of electrical lighting in theatres, enabling the late nineteenth and early twentieth century Symbolist and proto-Expressionist designs of Adolphe Appia (with his concern for "abstraction.... within the context of a three-dimensional sculptural setting that unify the performer and the stage space" [Collins and Nisbet, 2010, p. 148]) and the abstractions of designer Edward Gordon Craig for whom "action, words, line, colour, and rhythm....

appeal to our eyes and ears by a dextrous use of these five factors" (Craig, 1911, p. 140).

Plastique Fantastique and magic powder

An ongoing confluence of magical working, live art, performance writing, dematerialization of sculpture and light show, in which diagrammatical concepts and patterning are overtly applied, is present in the performances of contemporary art collective Plastique Fantastique. In their *All The Fantasies Of The People!* (see *figure 2.5*), a performance by the group at the IMT Gallery in London in 2013, there were both immersive and estrangement techniques at work. These included intertitles, psychedelic lighting, drone music that flattened rhythms, burned incense, a wall poster of a cone diagram, stills, and video projections. The audience was not allotted space – the inference being that the performance was for the benefit of some presence other than the spectators – but were clustered around cuboid table structures on the lower and upper surfaces of which planetary symbols had been laid out in something like sand mandalas (echoing Henry Daniel's layers and the *vévés* of Vodou).

After the upper mandala was swept away, powders and gooey substances were applied to the body of a reciter. Other than the reciter, the performers wore masks that were part-corporate, part-occultist, and part-terroristic balaclava. Avatars were invoked by the reciting of (apparently automatically-written or channelled) texts from a viable "P. A. Kingsford" and "Roy Burns" via the super-heroic Papa Mao and Voidrider to the trash magic of Subkast Kofkee and, finally, the folk-horror figure of Twigglett who manifested on video with their own dancing rave-avatar at a footpath crossroads around a grass triangle, only for this "character" to be embodied live, in mask and costume, by the reciter. Accompanied by frenetically building percussion and swiftly intercut video images, with traffic racing by outside the gallery door, the reciter became "live Twigglett," present and outside of time, dressed in sticks and a folkloric mask. "COMMUNIQUE ENDS" read an intertitle, followed by a video of ravers "jacking"; subjectivity was being ritually subjugated to the beat.

The use of materials by Plastique Fantastique, often deploying glitter, is ambiguous. As with Crab & Bee's *Scoriton Mysteries* (see below) – in which flying saucer sweets might be communion wafers, and black treacle might be a component of aviation fuel – the powders used in *All The Fantasies Of The People!* might be powder paint, or they might be magic dust. Which quality is not determined by circumstances or uses, but instead such possibilities are inherent qualities of the materials.

This approach to materials follows that of the anthropologist Martin Holbraad, who describes in his essay *The Power of Powder* (2007, pp. 201–217) his realization that the magical quality of *aché* powder used by Cuban *Ifá* diviners was not understood by them as due to how it was prepared or what

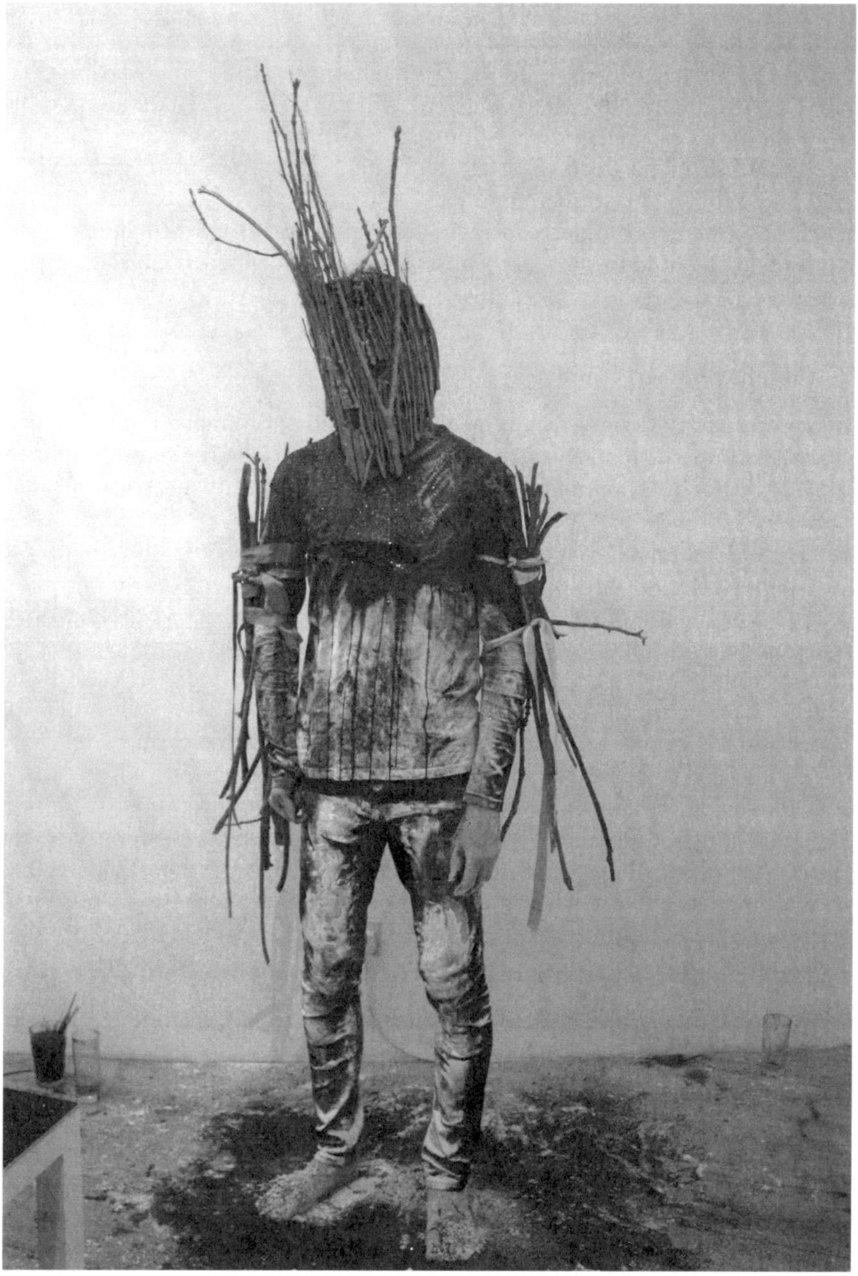

Figure 2.5 *All The Fantasies Of The People*, Plastique Fantastique, at IMT gallery, 2013. Image: IMT gallery.

effects it had. It could have these properties, but such an "understanding" of them would thereby separate the *aché* from its qualities. Instead, for the *Ifá* practitioners, the properties of *aché* (including "grace," "magical power," "all the powers, life, the secret of the earth forces" [p. 201]) are inseparable from the powder itself; power is powder and powder is power. To put an interpretation in between the two undermines any attempt to grasp what *aché* is. Similarly, Plastique Fantastique's use of materials indicates that the crucial quality of the objects in their (diagrammatical) performance is their contradictory but hybrid nature – abstract and creaturely – which is compatible with the dynamics of diagrams.

Post-dramatic diagramming

Given that drama is essentially a doing of (mostly human) relations, diagrammatical performance is necessarily post-dramatic in the sense characterized by Hans-Thies Lehmann as a "de-hierarchization of theatrical means" (2006, p. 86) and "the decomposition of the human being" (p.163) in a dynamic that "balances between a metamorphosis into a dead exhibition piece and.... self-assertion as a human being.... returning to things their value and to the human actors their experience of 'thing-ness'" (p.165). Echoing something of the idea of a thing's "withdrawal" from Object Oriented Ontology – "irreducibility downward to its components.... irreducibility upward to its effects" (Harman, 2019, p. 2) – what is most important here is the stubborn materiality of thingness irrespective of its relation to performers, or how things are interpreted or used; the powder flies up, then settles, the treacle sticks, and the flying saucer sweets are papery. As anthropologists Amiria Henare, Martin Holbraad, and Sari Wastel have argued, "meanings are not 'carried' by things but just *are identical* to them" (2007, pp. 3–4). In that way, the materials of a diagrammatical performance, perhaps surprisingly, are less symbolic than defiantly withdrawn, by their weighty presence and recalcitrance they nail down the abstract diagram to its creatureliness.

It is beyond the range of this book to judge just how widespread has been the use of diagrams in performance, or how common they have been a part of any assemblage of patterns or restorative "salvagepunk" in other artforms (see Salvagepunk section in Chapter 5). Nevertheless, the examples given above should be enough to suggest that they were and are present in sufficient impressive practices to warrant the claim that their use, relevance, and affordance in contemporary performance constitute some kind of continuity or debt even when not in a direct line of tradition and transmission. We are not arguing for any common tradition or genealogy of diagrammatical performing, but rather that there have been and are numerous constituents, with varied histories, of the kinds of diagrammatical performance that we are practicing and proposing. While none of the above performances alone provide all the qualities necessary for what we are proposing here (and equally we have not knowingly adopted any of these practices, or the parts thereof), they have all – or their equivalents – contributed something to both our understanding and our practice.

References

ABC Arts (2024) 'Fulu Miziki: The Afrofuturist collective who make their own instruments'. Art Works, [online]. Available at: https://www.youtube.com/watch?v=dwl5DsRmFFE&t=115s. Accessed on: 15.1.2025.

Auslander, P. (2008) *Liveness*. New York & London: Routledge.

Ball, M. (2000) 'Sacred mountains, religious paradigms and identity among the Mescalero Apache', *Worldviews*, 4(3), pp. 264–282.

Braziel, J.E. (2016) 'Atis Rezistans (Artists Resistance): Voodoo street sculpture at the Grand Rue, Port-au-Prince', *Callaloo. Johns Hopkins University Press*, 39(2), pp. 419–437.

Collins, J. and Nisbet, A. (2010) *Theatre and performance design: A reader in scenography*. Abingdon: Routledge.

Coulson, S. (2008) 'Response to "World's oldest ritual site? The Python cave at Tsodilo Hills, world heritage site, Botswana"' by Lawrence H. Robbins, Alec C. Campbell, George Brook and Michael L. Murphy'. Nyame Akuma, no. 26, December, 2008, pp. 2–3.

Coulson, S., Staurset, S. and Walker, N. (2011) 'Ritualized behavior in the middle stone age: Evidence from Rhino Cave, Tsodilo Hills, Botswana', *PaleoAnthropology*, 2011, pp. 18–61.

Craig, E.G. (1911) *On the art of the theatre*. London: Heinemann.

Daniel, H. (n.d.) 'The Touched Project: 2007-2010'. Available at: https://www.henrydaniel.ca/the-touched-project. Accessed on: 7.2.25.

Daniel, H. (2022) *Re-choreographing cortical and cartographic maps*. Bristol: Intellect Books.

David, Bruno, Russell Mullett, Nathan Wright, Birgitta Stephenson, Jeremy Ash, Joanna Fresløv, GurnaiKurnai Land and Waters Aboriginal Corporation, Jean-Jacques Delannoy, Matthew C. McDowell, Jerome Mialanes, Fiona Petchey, Lee J. Arnold, Ashleigh J. Rogers, Joe Crouch, Helen Green, Chris Unwin and Carney D. Matheson. (2024) 'Archaeological evidence of an ethnographically documented Australian Aboriginal ritual dated to the last ice age'. *Nature Human Behaviour.* Available online at: https://www.nature.com/articles/s41562-024-01912-w. Accessed on: 7.3.25.

Davy, K. (1978) 'Richard Foreman's ontological-hysteric theater: The Influence of Gertrude Stein', *Tulane Drama Review*, 24(1), pp. 108–126.

Gordon, Leah (director) The Sculptors Of The Grand Rue (video), 2008.

Harman, G. (2019) *Art + objects*. Cambridge & Medford: Polity Press.

Hebblethwaite, B. (2012) '*The cycles of the Voudo Ceremony*'. Posted 12.3.12. Temple University Press. Available at: https://templepress.wordpress.com/2012/03/12/the-cycles-of-the-haitian-vodou-ceremony/. Accessed on: 11.1.25.

Hebblethwaite, B. and Weber, M. (2016) 'Arabian Religion, Islam and Haitian Vodou', in Celucien L. Joseph and Nixon L. Cleophat (eds.) *Vodou in the Haitian experience: A black Atlantic perspective*. Lanham & Boulder: Lexington Books, pp. 209–239.

Henare, A., Holbraad, M. and Wastel, S. (2007) *Thinking through things*. New York & London: Routledge.

Holbraad, M. (2007) 'The power of powder: Multiplicity and motion in the divinatory cosmology of Cuban Ifa (Or *Mana*, again)', in Amiria Henare, Martin Holbraad and Sari Wastel (eds.) *Thinking through things*. New York & London: Routledge, pp. 189–226.

Kirby, M. (1973) 'Richard Foreman's ontological-hysteric theater', *Tulane Drama Review*, 17(2), pp. 5–32.
Lachman, G. (2018) *Dark star rising*. London: Penguin Publishing.
Lehmann, H.-T. (2006) *Postdramatic theatre*. New York & London: Routledge.
Malina, J. (2012) *The piscator notebooks*. New York: Routledge.
McCool, A. (2022) '"We like the sound of PVC tubing": Fulu Miziki, the band who make music out of trash'. *The Guardian* [online]. Available at: https://www.theguardian.com/global-development/2022/jul/29/fulu-miziki-congolese-band-music. Accessed on: 15.1.25.
Mirčev, A. (2024) *The poetics of performance diagrams*. Cambridge: Cambridge University Press.
Montréalserai. (2021) 'Wamunzo: Conversations With Zab Maboungou and Elli Miller Maboungou'. Posted 24.10.21. Available at: https://montrealserai.com/article/wamunzo-conversations-with-zab-maboungou-and-elli-miller-maboungou/. Accessed on: 7.2.25.
RISING. (2023) 'Talks and Performances at Ghetto Biennale'. Available online at: https://rising.melbourne/wormhole/talks-and-performances-at-ghetto-biennale. Accessed on: 11.1.25.
Schwarz, M. (2017) 'Rhythms without end – Haitian Vodou drum music'. Lakou Brooklyn. Originally published (2003) in *World Percussion & Rhythm* Magazine. Available online at: https://www.lakoubrooklyn.com/blog/2017/11/22/rhythms-without-end-haitian-vodou-drum-music. Accessed on: 7.3.2025.
Smith, D. (2011) 'A ritual and incantation-prayer against ghost-induced illness: Shamash 73', in Alan Lenzi (ed.) *Reading Akkadian prayers and hymns*. Atlanta: Society of Biblical Literature, pp. 198–201.
Smithson, R. (1996) 'A provisional theory of non-sites', in Jack Flam (ed.) *Robert Smithson: The collected writings*. Berkeley & Los Angeles: University of California Press.
Sullivan, S. and Low, C. (2014) 'Shades of the rainbow serpent? A KhoeSan animal between myth and landscape in Southern Africa – Ethnographic contextualisations of rock art representations', *Arts*, 3(2), pp. 215–244.
Tohonnie, J.J. (2012) 'About Apache Dances'. PBS Learning Media. Available at: https://www.pbslearningmedia.org/resource/natam.arts.dance.abapache/native-american-culture-about-apache-dances/. Accessed on: 12.1.25.
Welburn, A.J. (1981) 'Reconstructing the ophite diagram', *Novum Testamentum*, 23(3), pp. 261–287.

3 Why is diagrammatical performance relevant now?

Climate crisis discourses and performance

Fungus or ash?

On a recent train ride into Cornwall (for one of Crab & Bee's daily walks), it was hard not to notice just how lush the green landscape was becoming with recent extra rain and warmer temperatures. Yet within each burgeoning tree line there were always the skeletons of dead ash trees. Rhododendron bushes were a constant trackside presence.

Invisible from the train was the thriving *Hymenoscyphus fraxineus* fungus that causes "Ash die-back disease" and kills many of the local ash trees. To the trees this fungus is unfamiliar; they have had insufficient time to develop a resistance to it since it began to appear in Europe approximately 30 years ago. Meanwhile, the ever-expanding reach of *Rhododendron ponticum*, a "garden escape" widely planted by Victorians, now threatens the few remaining fragile pockets of temperate rainforest in the Western parts of the UK.

To what extent were we witnesses to – or enabling participants in – an unnatural process of species distribution? Are humans unnatural? If not, then is the intentional import of *Rhododendron ponticum* and the accidental introduction of *Hymenoscyphus fraxineus* simply parts of a natural process? Why should we judge the ash tree (*Fraxinus excelsior*) as more worthy of proliferation than the *Hymenoscyphus fraxineus* fungus? Is the Rhododendron's capacity to "wipe out vast swathes of native woodland plants.... Particularly troublesome in our temperate rainforests" (Bond, 2024, unpaginated) an unnatural threat or a part of a natural process of change?

Discussing the eradication of black and brown rats (*Rattus rattus* and *Rattus norvegicus*) from the Island of Lundy between 2003 and 2006, done in order to restore breeding populations of sea birds, Caitlin DeSilvey, Simon Naylor and Colin Sackett argued that "the cull rested on a set of moral decisions: that introduced species should be removed to support indigenous species; that less charismatic animals should make way for more popular ones" (2011, p. 13). Yet, are we right to liberally welcome and protect the diversity of plant and animal arrivals – making "moral

DOI: 10.4324/9781003606673-4

decisions" based on human equivalents – when the consequences of that might be a reduction in the kinds and number of species in the arrival area? And what do we mean by "consequences," when complex systems of predation and infection have been the motors of previous diversifications and we have little means of judging or predicting what any medium and long-term consequences will be, nor describing the dynamics of how any changes emerge?

In those cases where an animal appears to be thriving as a result of the effects of human industry or pollution (cockroaches, lice, tics, rats, mice, herring gulls, pigeons) they are often characterized as parasites feeding off the misfortune of other species or as collaborative perpetrators in a calamity. For example, changes in the acidity levels in oceans – deadly for molluscs – favour most jellyfish. Rather than celebrate the jellyfish's good fortune, the proliferation of these creatures (low in the hierarchy of nature when defined by fixed nostalgic views of a "good" or "natural" past) particularly around the in-pipes of the cooling systems of nuclear power stations, gives rise to dystopian visions of a return to the jellyfish domination of the Pre-Cambrian, Ediacaran oceans: "no spectacular coral reefs. No vast filtering mussel beds. No sharks slicing through the water.... Just jellyfish" (Gershwin, 2013, p. 344). Characterized as symptoms of eco-system degradation and long considered as "trophic dead-ends" (given that jellyfish are supposed to be eaten by very few other animals in the sea), this representation is now under threat, as "new technologies are revealing just how widespread jellyfish consumption is among marine predators" (Hays, Doyle, and Houghton, 2018, p. 878).

Apocalyptic traps

One of the drawbacks in assuming an apocalyptic climate narrative, is that humans get into a competition with more-than-human others for the role of most important victims. Given that more-than-human animals appear almost exclusively as losers and victims in our climate change narratives, this competition threatens to entirely remove those animals from meaningful discourse.

In climate change there can appear to be "winners": animals that are adapting far quicker to change than others (in particular, humans). One example would be the Turks and Caicos anole lizards. Those Turks and Caicos lizards that have survived the most recent hurricanes, much greater in force and intensity than before, have been those ones with larger toe pads and longer front legs, which allow them to hang onto plant stems when other lizards are swept away (Hanson, 2022, p. 120). In a brutal demonstration of "natural selection," it is the DNA of these longer-limbed lizards that is now being passed on. Such adaptation is not a novel process, but part of the ongoing evolution of species that never stops (matter in motion) and – as the fossil record demonstrates – is characterized by the sudden evolutionary spurts of a "punctuated equilibrium"; an adaptation of adaptation.

These Turks and Caicos lizards are, of course, only apparent "winners." There are rarely, if any, actual winners. For not every adaptation serves a species well. The fence lizard, for example, has adapted to global warming by spending more time in the shade, but this has reduced its time for feeding and mating (Hanson, 2022, p. 49), putting stress on the species. Given that nature is a web rather than a foot race, a species "out-competing" other species may soon find its own food sources disappearing as a result; as other less "successful" strands drop away the "winner" begins to slip through the web, like the overly-efficient parasite that kills its host (the Pre-Cambrian domination of the seas by jellyfish did not last forever). So, our attention to the movement and change of species is not a celebration but a recognition of an evolution specific to climate change, and how it will alter both the identity and, significantly, the relative dynamics among our more-than-human collaborators, and between us and them. Previously, humans considered it entertaining (and found it lucrative) to train large mammals – by punishment and reward – to participate (mostly obediently) in our cultural and sporting performances. However, in the near future a far more demanding, amorphous, volatile and higher-status cast of viruses, algae, fungi, jellyfish blooms and insect swarms is going to step forward.

How will we attend to and perform with these novel biological agents?

Stages and landslips

Similarly, assumptions about the stability of theatres and other performance buildings will be called into question. On a global scale, the frequency of earthquakes does not appear to be increasing at present (British Geological Survey, 2010, unpaginated), though recent studies of the effects of glaciers melting suggest that the lessened suppression of fault activity as a result will increase earthquake activity in the near future (Hurtado and Gallen, 2024, unpaginated). What, however, has *already* increased is the rise in sea levels challenging the viability of many island communities – "in the Solomon Islands and Micronesia, the islands of Hetaheta and Sogomou have lost 62% and 55% of land compared to 1947, while Kale and Rapita have been completely lost" (Vousdoukas et al., 2023, p. 1554). Also increasing is the risk of flooding worldwide due to global warming's heavier rainfall (Met Office, n.d., unpaginated). Wetter ground is less stable and more susceptible to landslips. What performers might have previously taken for granted as "safe" and stable spaces, free from rainfall, traffic noise, changing sunlight, and so on, may come to seem far more porous, and open to volatile geological agents, just as local biomes are welcoming unfamiliar arrivals. So, how will we attend to and perform with such novel geological and climatological agents, just as we already perform in and with the constant presence of microplastics (Li et al., 2024)?

In 2019, in our city of Plymouth (UK), composer Nuria Bonet created a duet for clarinet and herring gulls (partly recorded and played from edited recordings) entitled *Queen Canute*. It was premiered at that year's Peninsula Arts Contemporary Music Festival. This was maybe the first public attention to the city's ubiquitous herring gulls and their adapting to urban conditions that was not a negative address of their scavenging discarded food (and sometimes taking sandwiches direct from people's hands). These birds have become urban scavengers since the depletion of the herring shoals by trawlers and factory ships, and, more recently, since waste incinerators replaced the open landfill sites of Britain. Bonet has no illusions about the lack of agency of the gulls in her performance, and their plenitude of agency in everyday life; she describes how she first dreamed of making the piece but "with disastrous results because of the seagull's unwillingness to attend any rehearsals." Bonet is clear that no gull, nor any animal, is ever operating as either a musician or a musical instrument in her composition (Bonet, 2019, unpaginated). Following Nuria Bonet's example, in 2024 *Gullfest*, Plymouth's first arts festival in celebration of the herring gull was held: a weekend of "art, moving image, performance, spoken word, music and ritual celebrating Plymouth's vibrant gull scene" (ButCH, 2024 unpaginated) organized by the city's 37 Looe Street Gallery, including workshops, a "Gull Choir," and a costumed procession through the city streets.

In 1969, in a work he entitled *La Pluie (Projet pour une texte)* (*The Rain [Project for a text]*), the Belgian artist and poet Marcel Broodthaers was filmed writing with pen and ink under a torrent of water, the rain repeatedly washing the ink across the page as again and again the artist dips the pen in the ink and attempts to write. (The film is available at: https://www.youtube.com/watch?v=EHhigM53ILw). The rain does not seem to disappear in the work of art; it is not appropriated or disappeared in words or still images but persists. Yet, because the action is filmed, it removes the rain by means of representation (which is necessarily temporary); the film comes to an end, although the actual rain cannot escape to any "magical place called *away*" (Morton, 2022, p. 46). Despite its celebration of the agency of the rain, the use of media representation in *La Pluie (Projet pour une texte)* poses a challenge to the materiality of things we have chosen as allies.

Material allies

While the nature of water makes for an overtly agentive ally, other less obviously active materials can be equally vibrant and reactive to both their environment and their co-performers. In some of her place-based artworks, the Cuban-American performance artist Ana Mendieta (1948–1985) put herself and other materials directly into the matter of a "site." In her *Silueta* (Silhouette) series (1973–1985), Mendieta placed her body into mud, snow, and sand along with other objects such as leaves, flowers, gunpowder, and grass, sometimes burning things in the indentations. She also made body prints on various surfaces, using mud, paint, and blood. As an exile from her native Cuba and living in New York, her intent was to reconnect herself: "I become an extension of nature and nature becomes an extension of my body.... a reactivation of primaeval beliefs.... in an omnipresent female force, the after-image of being encompassed within the womb" (Mendieta, quoted in Barreras del Rio and Perrault, 1988, p. 10). When Mendieta steps out of the indentation or impression that she has made, the materials left behind continue to work and play; they burn, sink, slide. In the subsequent images of these artworks, if Mendieta is present she is also appropriated and assimilated by these other matters; caked in mud, foliage, or wax. Strangely, in the still images recording these actions the materials do not escape their representation (as the rain falls off the screen in *La Pluie [Projet pour une texte])* but persist despite their representation.

When approached from such a perspective as the examples given above – attending to agentive, newly arriving, and changing more-than-human agents – some of the prevalent nihilism and feelings of anxiety and inadequacy in the face of climate catastrophe discourse fall away. A redistribution of attention, from human destiny to something more ecologically global and philosophically flattened (similarly experienced by many people involved in rewilding [Wynne-Jones, 2022, pp. 3–4]), relieves the illusion of the obligation of husbandry (dominant agency and global responsibility) from humans.

Despite a common "green" rhetoric of "nature betrayed" or of humans losing their instinctive understanding of nature, there is no evidence for any fixed state of balance or defined diversity to which contemporary situations might return. In more-than-human nature, there is no absolute "native" (see Bond, 2024, above) or "indigenous" character (see DeSilvey, Naylor, and Sackett, 2011, above) to which a local biome can conform or be returned. In complex systems like biomes, there are no credible states that are ever "at rest" (to draw a notion from Newtonian physics). Various candidates have been brought forward to play the role of such an ideal fixed state in the past; conservationists have proposed a preservation of, or return to, a mostly imaginary eighteenth century pre-Industrial Revolution English landscape (DeSilvey, Naylor, and Sackett, 2011, pp. 31–32), "Merrie England" radicals have argued for a return to a medieval "closeness" to the land (Blatchford, 1894, pp. 19–23), and "back to nature" movements have advocated a primitive communism ("communes") of hunter-gathering and small-scale farming. However, modern archaeology has revealed just how far back eco-destructive human practices go, with vast projects of deforestation and burning perpetrated by our Middle Stone Age ancestors (Brown, 1997, pp. 134–139). There is no recent or ancient utopian Edenic model to which to aspire, or against which to judge the present, and no eternally valid eco-principles to fixate upon as guidelines for orienting our judgements. The absence of which is not a reason for the flagrant destructiveness of an accelerationist "drill, baby, drill," but rather an indication of the complexity and longstanding nature of the problem.

Pharmakon

Crab & Bee, unattracted to either dreams of a return to balance or recourse to angry destructiveness, have been driven more by our ongoing dialogue (see "Sexy Theory" below) around all these subjects; and by alliances with the matters under threat in these circumstances. If there is any principle guiding our overall approach, it is the idea that "matter and motion [are] ontologically primary and not derived from some less material principle" (Nail, 2024, p. 2). We have not looked to Gaia, Platonic ideals, a lost ecological "golden age," or an intelligent designer. Instead, we have arrived, via "process" philosophers like Henri Bergson, Gilles Deleuze and Felix Guattari, at something approximating to Thomas Nail's kinetic materialist approach to "self-caused" (p. 40) matter.

Influenced by ideas from sources as varied as Lucretius's concept of the swerve in matter, Marx's early thought, Virginia Woolf's depictions of consciousness, and quantum mechanics, Thomas Nail describes matter as indeterminate, always in motion, relational, and never separate from its processes. Long before we were to read it in Nail's work, we had embraced a resistance to any suggestion, inherent in the idea that matter could ever be inactive or part of "a natural hierarchy in which some things, people and ideas are.... passive material to be manipulated and mastered" (p. 6). Rather, following Jamie Mcphie's idea that "concepts are performative" (2018, p. 308), we have long assumed that all things, including concepts and theories, are material agents in motion and with which we are in a trialogue.

It would be unwise to attempt to present here a coherent, singular, eco-theoretical basis for diagrammatical theatre, as theorization is always in danger of sealing itself off from a messy ecological meshwork. Instead – scrambled together from "being there" with abject terrains, a grasp of the mechanics of a "society of the spectacle" which all the time seeks to reduce things to their representations (Debord, 1994, pp. 12–24), and our repeated indulgence in new vitalist process theory – we offer what Nail, borrowing from the ancient Greeks, calls "a *pharmakon*: a practice of steering clear from poisons and experimenting with potential remedies" (p. 8). At times we will adopt a homeopathic approach to such "poisons"; avoiding ingesting a fatal dose while attending to the effects of, say, a toxic site in order to imagine "potential remedies."

The poisons come in various forms. For example, global human political attempts to ameliorate or reverse the effects of climate change are often advocated in terms of how they might benefit the very economic and industrial systems that have created climate crisis in the first place. The clues to such confused practices are often held within the language used to present such programmes. Medecins sans Frontiers, for example, put forward a plan to reduce "their environmental footprint" whilst addressing the needs of "vulnerable and excluded populations" in a time of accelerating "climate change and environmental degradation" and called this plan – with no hint of irony – "Environmental Roadmap" (Medecins sans Frontiers, 2022). In 2024, the DeGroote School of Business, McMaster University, Hamilton, USA hosted an online webinar "addressing complex aspects of sustainability, exploring its significance, challenges, and practical solutions" (2024), entitled – again with no hint of irony – "The future of sustainable practices: paving our way to a greener future" (DeGroote, 2024: online).

In some cases, these representations are nothing more than shadow versions of reality; like the exposure of organizers of both the 2023 COP 28 and 2024 COP 29 summits (for world leaders addressing climate change) as using the summits to promote fossil fuel trading and advance the interests of eco-catastrophe. We might consider these representations as simply manifestations of the self-interest of heinous owning classes and their facilitators, willing to trade long-term general depredation for short-term profits. However, in an increasingly post-industrial information society, in which the overlap of the being of a thing and its representation is growing ever more exact (Debord, 1994, p. 12), the entanglement of annihilation and image is increasingly tight. So, prising the thing from its representation, as part of a performance – in *Scoriton Mysteries* (2025) our returning of engine parts (represented as parts from a UFO) to the status of trash – seems to work something like a representational equivalent of re-wilding. Equally so, the enactment of the loss of trees (by sprinkling wood ash) around a new housing estate called "Badger's Wood" built over former woodland.

From images to matters

In all of these discourses, whatever the political positioning they emerge from, the viewpoint is – almost without exception – human. If more-than-human species are addressed, it is mostly in terms of a threat to humans – for example, colony collapse among bee populations endangering the pollination necessary for the propagation of human food supplies (Stankus, 2008, unpaginated) – or as victims of human-driven changes damaging animal populations. Such endangered animals, as they disappear in reality, increasingly figure as symbols of human hubris, reinscribed as containers for the effects of human wrongdoing:

> Zoos, realistic animal toys and the widespread commercial diffusion of animal imagery all began as animals started to be withdrawn from daily life. One could suppose that such innovations were compensatory. Yet in reality the innovations themselves belong to the same remorseless movement that was dispersing the animals.
>
> (Berger, 2009, p. 27)

The use of images of polar bears "stranded" on a melting ice floe and the move away from them as adequate representations of climate change has been telling (O'Neil, 2022, unpaginated); given that polar bears have always encountered a seasonal melting of ice and are capable of swimming distances upwards of a hundred miles. The reality of both the polar bear and climate change disappeared in the telling of these images. Rather than explaining the nature of the threat to the polar bears' habitat, with its causes and effects, such representations have occluded it; they have become a part of the disappearing of the polar bear. The transformation of matter into representations has been coterminous with the obliteration of matter; similar to how a world, once divided into individual material commodities for exchange, can be allotted an economic value that is independent of its use-value – the "sensuous qualities" of things being used – and reducible to its non-material representation (see Nail, 2024, pp. 53–54).

The existential threat of climate change problematizes existence; it begs questions of the nature of being (that of ours and of others), the nature of "sensuous qualities" and their capacities and limitations for knowing other existences, and of the existential relevance – if any – of identities, separations, and distinctions. If we assume, as the arguments above imply, that an indeterminate existence is (necessarily) performed, rather than something that is given by a god or by the laws of physics, then performance itself hovers somewhere between representation and the conservation/change of the thing itself.

Performance shapes a swerve at the same time as celebrating that swerve as being what performance is. So, rather than our performances being *about* matter in a time of climate change, if we take seriously the implications of our thinking, then in our performances we are seeking after a kind of matter, a manifestation of swerve, an indeterminate dynamic of unplanned change generated within the interaction of the various materials we bring from a porous and unbounded place to a non-site defined by separation. In diagrammatical terms, we create a funnel for things, with its mouth in the place of "being there" and its tip of condensation in "the three-dimensional metaphor" of the non-site (Smithson, 1996, p. 364).

Transhumanism, transcendentalism, and diagrammatics

Cultural trends

The three great twentieth-century global projects – neo-liberal capitalism, anticolonial nationalist struggle, and socialism – massively increased production while fostering total war, spectacular media, racism, bouts of fascism and genocide, and accelerated the oncoming rush of ecological catastrophe. At the same time they generated, unevenly, some significant improvements in global human health care and standards of living, yet still increased immiseration for many. An attraction to various kinds of anti-materialist philosophy and practice (mysticism, apocalypticism, theocratic beliefs, psychedelics, ecstatic religions, cybernetic futurism, the pursuit of the singularity) is, then, hardly surprising. This tendency has become more urgent and frantic in the second and third decades of the twenty-first century. Partly this is due to the high-profile projects of tech billionaires seeking technological fixes to both climate crisis and the human condition. Their proposed and active projects include terra-forming other planets (exporting ecological crisis), bio-engineering human immortality, and uploading human consciousness to AI-based computers in the form of digital information.

These trends have also been fuelled in the fields of philosophy and critical theory by moves towards transhumanism, anti-humanism, and dematerialization. Influential in this was the thinking that emerged from the activities of groups like the Cybernetic Culture Research Unit at Warwick University in the 1990s. Their discourses were as diverse as the "capitalist realism" and hauntology of Mark Fisher (2014) and the early versions of the "dark enlightenment" of Nick Land (2022). They built upon existing movements in cultural practice, such as the turn to performance in a dematerializing fine art in the 1960s onwards (Lippard, 1973), while many progressive, feminist and experimental artists and thinkers were drawn to the celebration of a blurring of distinctions between animals, humans and machines, as imagined in Donna Haraway's 1985 essay *A Cyborg Manifesto* in the *Socialist Review* magazine.

Neither transhumanism nor transcendentalism – transforming ontological expectations in order to become things (immortal, immaterial) that humans are not – are necessarily modern movements. While such aspirations are specific to many contemporary religions, secular futurisms and are now part of the "conspicuous consumption" of the super-rich (Galbraith, 2001, pp. 210–211), they have not always been exclusively so. In number 139 of the *Pensées* of Blaise Pascal, first published in 1670, the philosopher argued that such aspirations are a symptom of the human condition itself. He describes the predicament of a gambler who, when offered all their possible winnings on condition that they cease playing card games, refuses the offer. This gambler also turns down the offer of continuing to play card games on condition that they play for nothing. Pascal's conclusion is that people play games (and by implication perform other kinds of risk-taking in life) not simply for entertainment; there needs to be something at stake in a game/life, and yet that thing at stake, when separated from the risk of the game/life itself, turns out to be of little or no interest to the person. Thus, according to Pascal, humans need to invent *imaginary* needs and intrigues, not to resolve them, nor out of a desire for any reward for doing so, but instead to have a reason to keep pursuing them (Pascal, 1958, p. 40).

What is different now, however, is the prospect of a "singularity"; the promise of a technological conclusion to this "game" of life. The stake is now the game itself. The "singularity" hypothesis was introduced and discussed among mathematicians and technologists in the 1950s and popularized from the 1980s onwards. It proposes a moment in the future when technological development, particularly in the field of artificial intelligence, arrives at a point where it removes all the imaginary needs and intrigues from the "game" as it passes beyond humans' capacity to control it. This will, so the argument goes, cause a spasm in human society; it collapses the condition of the human/gambler, as presented in Pascal's *Pensées,* as its transhuman transformation is produced by a non-imagined thing-at-stake (the all-powerful singularity of the technology) which is as indifferent to its effects on the gambler as the gambler is to it.

The posthuman virtues of diagrammatics

One of the useful qualities of diagrammatics is how, given its indifference to human failings or virtues, it pre-empts transhumanism by offering a posthumanist indifference to the human condition without requiring a transcendence of it. For diagrammatics is not conducive to a human drama, even an apocalyptic one. Another of its virtues is its close attention to structural dynamics, exposing any recourse to existing ideological patterns, even in the futuristic aspects of the idea of a singularity, which mirror certain patterns in theological eschatology.

Despite such wrinkles, the idea of a singularity – an abstraction and an indifference in a chance meeting within a mainframe – poses an insidious threat to the "creaturely" qualities of diagrammatical performance. The markers, geometrical planes, and abstract ideas of a diagram can attract a way of thinking – such as the model of a singularity – that dislocates it from its connections with the cosmos and floats it free from our more-than-human allies, against which circumstance it needs to guard.

To prepare against this, a diagrammatical practice needs not only to be an embodied one, but to be based on identifying what is materially, rather than what is imagined, "at stake." It also requires attending to whatever that is, by repeatedly "being there" with it. (In the actual games Crab & Bee play, nothing is at stake; they are there only to demonstrate diagrams in action.) For that reason, many performances that deploy diagrams (including our own) eschew the common naturalistic dramatic dynamism of having, like Pascal's gambler, an invented "something at stake" (examples being, in "Doctor Faustus" [1594] the soul of the main character, in "A Midsummer Night's Dream" [1595] the happiness of various couples, in George Bernard Shaw's "Pygmalion" [1913] the identity of Eliza Doolittle). Instead, the "what is at stake" of diagrammatical performances tends to lie somewhere (a point) at the end of, or at a node along, the length of one of its connective lines, or across a conceptual field, all of which may be very distant from the place of performance and its performers.

In the idealist diagrammatics of Symbolist drama what was "at stake" was a connection to some otherworld of dreams and spirits. In a materialist diagrammatics – like that, say, of Peter Weiss's play *The Investigation* (1965) in which the testimonies of victims and perpetrators at the Auschwitz-Birkenau extermination camp were arranged in accordance with the circles of hell of Dante's *Inferno* (1321), moving from the arrival ramp to the gas chambers and ovens – the connections "at stake" were historical events and material processes. In our *Scoriton Mysteries* (2025) what is "at stake" seems at first to be the possibility of aliens arriving on Dartmoor from afar, but as we introduce the various shapes of the performance's objects, it turns out to be the nature of the hills of the moor.

Subjectivity, materiality, and tradition

An exemplary performance

On the 21st of January, 2023, a protest march of more than three thousand people against a ban on free wild camping on an area of Dartmoor (Devon, UK) took place. The protest started at the village of Cornwood, progressing to the area under threat on Stall Moor. We attended this walk up to the moor to show support for the protest. We were surprised by its exemplary performative nature. In notes made by Phil on the same day, he writes of the event:

> wonderfully diagrammatical – the disparate groups and individuals gathering in the village square, forming a head, then a snake up to the moor, dispersing a little and then re-forming the head again, to listen to an invocation of Crockern (by storyteller Martin Shaw) and to make a collective declaration ("we are Crockern!") – and then Crockern appears!

Having processed along the narrow lane up to the moor, the walkers congregated on the hill. Then, from a different direction, far away from the marchers, the spirit-monster appeared over the brow of a tor.

"Old Crockern" is generally accepted, locally, as the "spirit of Dartmoor": "[F]or centuries, Old Crockern, the spirit and guardian of the moor, has ridden out from Crockern Tor at night." Usually imagined as "a ghoulish figure of varying gaunt descriptions who rides a skeletal horse," Crockern is particularly pertinent to a protest around land rights as "[I]n years of enclosure the people called upon the spirit to resist the impeding of their common land." The cursing of the 2023 "land grabber" in the name of this spirit was also historically grounded: "Old Crockern's curse is powerful: there are stories of landowners attempting to farm and fence off the moor only to be forced out by waning financial circumstances or personal failings" (Fulford, 2023, unpaginated).

The wooden frame and stretched fabric of the Old Crockern figure (see *figure 3.1*) at first glance and at a distance gave it the appearance of a vegetable beast, or enormous turnip, like one of the alien monsters that emerge from the domes in the Hammer movie *Quatermass II (1957)*. As it approached, however, a strange grey head fringed with long multicoloured streamers became more distinctly evident. The design was enigmatic, limbless, unlikely. In Phil's notes, he writes: "is it a horse, is it a dog?… its ambiguity allowed space for 'us' to be it… and it stands just a

Figure 3.1 Old Crockern protest, Dartmoor (UK), 2023. Photo: Lisa Schneidau.

little higher than the crowd so everyone can see it, while the crowd moves around it, gathers, follows it, dances, shouts and claps." All the time, a small group of ribbon-clad "teasers" and musicians were creating a turbulence around the wood and canvas vegetal beast-god. The Guardian newspaper online described all this as a "theatrical ceremony" (Wall, 2023, unpaginated).

While there was no formal dance, it was clear from observing how comfortable the crowd was in moving around Crockern that there was some familiarity with the varieties of morris dancing practiced around the moor (including traditional, Beltane Border, and acid morris). In other words, unspoken patterns were at work, which were diagrammatical in the sense that although no political or religious identities were on show, the evidence of symbols and attitudes indicated patterns consistent with a variety of animist and modern-traditionalist loyalties.

Finally, individuals and groups broke away and re-formed the snake back down from the moor and to the village, where it dispersed. Given that this was a protest march with a simple demand, there were unusually diagrammatically-imagined aspects additional to the pattern of gathering, assembly, snaky march, gathering and invocation, feast/picnic, arrival of Crockern and dance, serpentine procession back, and dispersal. These were not due to the speeches in the village square or on the moor, or the expression of an organizing philosophy, but a result of disparate individuals and intergenerational social groups communicating to each other (and with Crockern) through the imagery of often humorous handmade placards ("SERF'S UP!"), dress, music, and collective movements. A quarter of a mile into the walk, we passed a cottage draped in twinkling lights and a banner that read "the stars are for everyone"; the residents fell into step with the protest, carrying large bright stars attached to hazel wands. A Romani flag and a huge pair of boltcroppers were waved above the heads of the crowd, as a walking tent was interviewed by the press.

Phil wrote in his notes that "there was space for the event to become itself." Despite the striking placard of artist Ethan Pennell (creator of *The Dartmoor Folklore Map*) which was based on his painting of Crockern as a pale, emaciated, skeletal figure with long thin limbs and clawed feet, the general uncertainty around the appearance of Crockern opened up a space for participants to imagine and invoke their own Crockern. So, rather than by recognizable materials or overt politics and philosophy, in counter-tendency to the general culture's shift towards dematerialization, the Stall Moor procession and ceremony were predominantly communicated in materials and embodied patterns. The event was "post-ideological" and pro-subjective, yet also collective, being based on exchanges within the event itself as well as beyond. Rather than various agents intervening to present more complex pre-thought-out ideas, demands, and strategies within the event (common on political marches), the event was working itself out. (In May 2025, the UK's Supreme Court upheld the right to camp on the moor without the landowner's permission. The performance worked.)

Non-instrumentality and subjectivity

The 2023 Stall Moor procession and "theatrical ceremony" blended together parts of two tendencies in recent performative protests around issues of ecology, climate change, and land. On the one hand, there was an element of trespass; while at the same time, care was taken to avoid confrontations that would allow police and private security to slip into their "comfort mode." On the other hand, there was a certain non-instrumentality about the event. This was in line with the development of the "Red Brigade" performances on Extinction Rebellion demonstrations in the years prior to the Stall Moor event. These red-clad and white-faced figures, moving through slow, choreographed gestures, often defusing points of conflict, expressed nothing precise – were they symbolic of fire, of blood? Were they ghosts, were they ecological spirits? – invoking whatever a spectator chose to intuit within the enigma of their concerned melancholia, implying a lack in urgent need of address.

In *The Three Ecologies* (2000), the psychologist-philosopher Felix Guattari placed a plateau of subjectivity on a par, and interwoven, with plateaus of social relations and natural environment; "an ethico-political articulation… between the three ecological registers (the environment, social relations and human subjectivity)" (pp. 19–20). He then subjected their relationships to a critique based on how historical changes (including progressive ones) had mostly not worked out how Guattari had hoped. It is within such an instrumentally sceptical and anti-nihilistic pessimism that a diagrammatical eco-theatre occasionally emerges, in tune with Guattari's proposal for an "unbalanced and unintegrated ecosophy." This is based on a refusal to rely on neither radical social change through collective activity nor a benevolent turn on the part of corporations. Instead, Guattari advocates for "ways of living" (Guattari, 2000, p. 20). He proposes an aesthetics based on imbalance – a "swerve" perhaps – with a relation of environment to subjectivity and subjectivity to environment as its motor. In the interactions of the Stall Moor "ceremony," in the enigma of the Red Brigades, and in the materials-based dramaturgy we are proposing here, the overarching diagrammatical approach is about making spaces for the agency of such subjectivities.

This means resisting anti-ecosophical pressures to concede to individualist practicalities (a technological fix, changes of individual lifestyle, new and better leaders or spokespersons) or to a transcendental "going over the heads of nature" to make a direct appeal to the divine. But what if we accept the subjective as part of a material cosmos? Could we embrace a subjectivity that reaches out beyond the individual, as the material senses reach out in James J. Gibson's articulation of their tentacular stretching beyond the body to find sensation and information (1966, pp. 3–5)? What if we do not concede to the nineteenth century's invention of an inner unconscious playing out in symbolic representations of experiences of a human-social past, but rather understand the unconscious and the conscious as invaded and permeated by the environment?

Learning

What if subjectivity and environment are not separated and discrete things? That may sound self-centred and individualistic, so it is important to always return to the "relation" part of Guattari's formulation. Attention to the materials of the terrain of performance is crucial to this dramaturgy.

At the same time, there is evidence to suggest that marginalized cultural agencies support such relations. For their 1995 book *Twilight Of The Celtic Gods*, David Clarke and Andy Roberts were able to track down a handful of individuals (all very protective of their identities) living in remote communities in the Derbyshire Peak District or the Yorkshire Dales (both UK) who claimed to be inheritors of longstanding family traditions of "learning" exactly such relations. In the words of one interviewee, who the authors call "the guardian," this "learning" received from his maternal grandmother from when he was seven year of age consisted of long walks, sometimes at night, "often in unfamiliar territory," sometimes in silence: "I began to realize what I was being taught. Slowly from being a series of roads, paths, fields and hills, I began to see the general topography.... in a very different way." Ignoring official signposts and "recognized rights of way," grandmother and grandson were "instead paying total attention to the land and how it was formed, its shapes, textures and hidden places." Rather than as the ground of a route to somewhere else, the land "was revealed to me as part of the body of a living thing whose mysteries were accessible if you looked in the right way, at the right time"; in other words, the child was being introduced to a creaturely diagram in the terrain. (Clarke and Roberts, 1995, pp. 22–24).

Such a "learning" is a horizontal one; it aspires to no command over its subjects, to no laws or conclusions, not even to theories or hypotheses. The "guardian" claims that there is "nothing supernatural" to any of this. It is an attention to material ground, reliant on the learner's subjective interpretation ("I began to see") of a living body with mysteries. Although the "night sky soon became populated with heroes and villains, gods and goddesses," there is no "worship" or ceremonial magic involved. Instead, "staying true to the tradition and marking the special times, being open to the natural powers was enough" (Clarke and Roberts, 1995, pp. 24–25).

Where Felix Guattari warns against "simply huddl[ing] around archaisms" (2000, p. 20), "the guardian" in Clarke and Roberts's book is indifferent to the longevity or cultural identity of his tradition ("[T]hat sort of thing is for others to work out" [1995, p. 24]). The tradition of "the guardian" is far closer to Guattari's "ways of living" than to either a nostalgic archaism or the "dreadful rise of religious fundamentalism" that the psychologist-philosopher identified as threatening "collective subjectivity" (2000, p. 20).

This is in common with many practitioners of "traditional" and "folk" arts and learning, who happily and self-consciously splice modern conventions (acid house or Goth style) to traditions such as wassailing and morris dancing, similarly to the elaboration of the costumes of the Crown Dancers. Or indeed, introduce new "old" traditions such as the Dark Morris "dusking," championed across meme culture and social media by academic folklorist Lucy Wright.

Dusking is the act of dancing the evening sun down on 31 October, as those in the Northern Hemisphere enter the dark half of the year. It offers a counter celebration to the Beltane Morris practice of dancing the sun up at Beltane, when the summer begins. The combination of traditional dancing with the global "nowness" of the internet connects people corporeally within a diagram of seasonal change, across the turning Earth they dance on. For as Lucy Wright suggests:

> [L]et's not forget the Morris community in the southern hemisphere, many of whom have long danced the sun down 'in solidarity' with their northern counterparts on May 1st, before dancing it back up again in October. Jack Frost Morris and Phoenix Morris from New Zealand and Hot For Joe Border from Australia, amongst others, have been Dusking in one form or another for many years [...] it seems that 'balance' is an important value to many established Morris sides, too.
>
> (Wright, 2023, unpaginated)

Times

The indifference of the "guardian" of *Twilight of the Celtic Gods* to the past may seem surprising to some, but it reflects an understanding of time as looping rather than linear. It is resonant with the project of those cultural planners and engineers in the documentary *Into Eternity* (director, Michael Madsen, 2010) who are tasked with designing the Onkalo nuclear waste depository in such a way that no one in the future will want to examine or explore it. Beginning the excavation of the ground to create the facility, the thought strikes them "what if we were to find a copper capsule now? What if a previous and now extinct society has been through all this before?" Similarly, the "learning" of "the guardian" is neither exclusively of past or present, it is a-cultural, it loops the past within the future and the future within the past, largely indifferent to traditional identities and authenticities – "[T]hat sort of thing is for others to work out" – and focuses on what David Burrows and Simon O'Sullivan describe as the "calling forth a people yet to come.... through technics of *looping* and *nesting*" (2019, p. 6).

Rather than the random slicing of William Burroughs's "cut up" (see Fictioning section below), a diagrammatical manipulation of time by its stuttering progress first repeats the fatal "separation" identified by Guy Debord as "the alpha and omega of the spectacle.... [which] makes no secret of what it is, namely, hierarchical power evolving on its own, in its separateness" (1994, p. 20). However, by then nesting fragments of time, one within another, and then looping sequences back on themselves, a diagrammatical dramaturgy can create connections without recourse to a reductive unity. Instead, like certain kinds of theory-fiction, it is "a writing from the future into the present" (O'Sullivan, 2024b, p. 14).

Paradoxically, it is the looping and repetitive nature of such practices that renders them available to marking changes of weather patterns, seasonal shifts, erosion, flooding, and so on. Such a nesting of fictions, one within another, "does something – recursively, as it were – to our own perspective on the fiction we move in and through (if only allowing us to see the latter as a fiction)" (O'Sullivan, 2024b, p. 13). Performing in a liminal space between the fiction of theatre and our fictioned reality – in a ritual zone where materials regain their materiality and become metaphorical – we can emerge as co-authors of a shared ecological terrain.

The cancellation of the future

The termination of modernity

In 2014, Mark Fisher called the opening section of his book *Ghosts Of My Life* "the slow cancellation of the future." This was less a vision of an apocalypse than a description of social malaise and cultural entropy. In all fundamental aspects, he argued, the future (as presently understood) could only weakly repeat what had already come before. Drawing mostly on European and UK conditions, Fisher based his social analysis on two premises. Firstly, that following the post-Second World War vibrancy and innovation – the rock and pop breakthrough of the 1960s (originating from the jazz and blues of Black America), "homemade" punk in the late 1970s, and the initial European rave culture of the late 1980s/early 1990s – much of Western popular culture had lapsed into a "fallow terrain" of weak repetition, sequel and imitation when "cultural time decelerated and went into reverse" (Coverley, 2020, p. 9). Secondly, that society was "living after the gold rush." The halting of utopian state-based social-reformism by the rise of neo-liberalism combined with successive economic crises had removed all prospects of a future appreciably different from the present. Worse still, this present was now leeching off the past: "culture has lost the ability to grasp and articulate the present.... there is no present to grasp and articulate anymore" (Fisher, 2014, p. 31).

Around the time of Fisher's death in 2017, popular angst and media anxiety about the depredation of the environment had shifted from occasional concern for certain species of birds, insects, or larger mammals to fears of a new global mass extinction event. The failure of shock capitalism to deliver either its promised stable, entrepreneurial and philanthropic world, or an equitable ("trickle down") distribution of affluence based on individual endeavour, had undermined belief in neo-liberalism. The election of authoritarians in India, Hungary, and Turkey, and the growing influence of the far right in Germany, France, and (most recently) the UK, was reflected during the 2016 US Presidential election, when numerous young conservatives publicly deserted libertarianism and neo-liberalism for organizations of the "alt right" (many subsequently "disappearing" into the cybersphere). Not only had the prospects of planned, regulated and state-protected post-war utopias been undone, followed by the falling apart of their neo-liberal nemesis, but the fundamental motors of the system (whether conducted liberally, democratically or dictatorially) now constituted a threat to its own existence. The modernist human project was in a terminal decline.

In *Ghosts of My Life (2014)*, Mark Fisher had responded to this termination of modernity. He refused to countenance postmodernity as permanent and inevitable, championing those contemporary artworks – like the music of Burial, Pole, The Caretaker and Belbury Poly – that deployed the processes of imitation and repurposing in order to keep alive "the tantalising ache of a future just out of reach." Fisher described how the squeaks and crackles of this music, "foregrounding rather than repressing sound's accidental materialities," were pushing back at the dematerialization of things.

Ontology/hauntology

The third section of Fisher's (2014) book is entitled "Hauntology." This is a reference to Jacques Derrida's influential cultural theory described in his *Spectres of Marx* (1993), in which he conflates ontology (the theory of being) with haunting. Derrida's purpose was to construct a theory that characterizes contemporary existence as a post-life (its future cancelled) that is experienced as spectral, more susceptible to images than matter. In articulating this ghostly landscape, Derrida brilliantly captured late capitalism's attempt to technologically transcend a rebellious and threatening Nature by shifting sideways into the controllable realm of virtual reality.

Turbo-charged by generative AI, plugged into digital geo-connectivity, these trajectories, identified by Derrida, Fisher, and others, have only grown in strength and reach. They have realized what the situationist Guy Debord identified in 1967 as a tendency in mass media society to generate "an immense accumulation of *spectacles*" by which a society "proclaims the predominance of appearances and asserts that all human life.... is mere appearance" (Debord, 1994, p. 14).

In Adam Curtis's TV film essays – *All Watched Over By Machines of Loving Grace* (2011), *Hypernormalisation* (2016), *Can't Get You Out of My Head* (2021) – the documentarian describes this transformation to an image and information society as a planned and theorized project. Deploying hauntological features – from the foreboding soundtracks of 1980s horror movies like that of John Carpenter's *The Fog* (1980) to eclectic sampling of promotional films depicting lost futures – Curtis identifies many of the key psychologists, intelligence agents, advertising gurus and cyberneticists who pioneered ways to order the consumption of images and ideas in mass societies. He describes how, in the last decades of the twentieth century, elected politicians surrendered much of their legislative, executive, and regulatory power to corporations and financiers, preparing the way for the internet's unregulated exchange of images. In such a rampant "society of the spectacle," the material world has not disappeared, but everyone in it is subject to "a social relationship between people that is mediated by images" (Debord, 1994, p. 12). The result, which Debord describes as "a world that really has been turned on its head" (p. 14), is every bit as radical as actual dematerialization: the subjection of things to their representations, and at its most extreme the replacement of things by their images.

This is something that Crab & Bee have repeatedly witnessed on our preparatory wanders: a housing estate, for example, called "Saltram Meadows" where the meadows are gone. While most trees have been cut down and all the badgers exterminated in another housing estate called Badgers Wood. An Oak Tree Close has no oaks. On pub signs and council notice boards, there is a preponderance of the images of animals – choughs, bears, wolves, red deer – in places where they have been hunted to extinction and/or their habitats destroyed (Billinghurst, 2019, pp. 56, 71–72).

The consequences of such dematerializations for embodied performance and the relation of humans to more-than-human others are clearly fundamental. Subjecting bodies to their representations (images which only have their being as parts of networks of media) is to deny them a space of "being there" in their individualized material agency. Where appearance dominates, live performance makers are denied their own subjectivity and agency; subjecting their artworks and representations of climate change to the demands of a spectacle. A storm becomes "a hoax," or framed by conspiracy theories about the seeding of clouds, or a punishment from God, as just natural payback for human excess, as an opportunity for "storm chasing" entertainment, or the revenge of an angry Gaia. There is little space left for the more-than-human material agencies of atmospheric pressure, precipitation, sea, wind, and rain.

Responses to the dematerialization crisis

Disembodiment

In December 2022, the *Performance Research* journal published an issue (Volume 27, Issue 8) dedicated entirely to articles "On Diagrams & The Diagrammatic." The coverage of deployments of diagrams or diagrammatical thinking included accounts of workshops, historical re-enactments, televisual (detective-style) pin-boarding, *dérive*, counter-mapping, and the charting of cultural practices in Portugal and of performance art venues in Switzerland. Yet, across the 165 pages of articles, accounts of contemporary performances (excluding the examples in Crab & Bee's article) were limited to cursory descriptions of Laura Paolini's *Make Your Bed* (2020) and Rubiane Maia's *Waterline* (2021), and three examples of interactive visual art.

Given the influence and apparent usefulness to academics of the *idea* of diagrams (sourced mostly via Deleuze and Guattari and Michel Foucault), it is perhaps understandable that an emphasis comes to be placed on thinking, theory, and notation over practice. There is a danger that diagrammatics, at least as represented in the *Performance Research* issue, could be seen as part of a more general disembodying of performance and as cultural dematerialization, vanishing the creatureliness. However, in the actuality of practice, diagram-influenced or inflected contemporary performance that knowingly deploys and responds to materiality, is, while not common, both widespread and significant. Examples include work by performance-makers such as Henrietta Baird, Plastique Fantastique, English Heretic, Miranda Whall, and Louise Ann Wilson (and others described below) as well as performative artists like Nolan Oswald Dennis.

Dematerialization of country

In response to the extensive fires of 2019/20 in Queensland, the Kuku Yalanji choreographer/writer Henrietta Baird, "working closely with her Elders" (Sydney Festival, 2025, unpaginated), created a four-part project *Plant A Promise* (2025) for the Bangaroo Reserve (Sydney, Australia). This included a dance performance, a Baya (fire) installation using burnt furniture from the fires, native planting sessions, and "yarns, which weave together Indigenous knowledge with environmental science to sharing the importance of caring for Country" (Blogg, 2025, unpaginated). Premiered on Gadigal Land in Sydney, *Plant A Promise,* was to later move along "the ancient saltwater pathway" (Bailey, 2024, unpaginated) to Kuku Yalanji Country.

Developed over a number of years, Henrietta Baird had begun *Plant A Promise* with a plan for collecting eucalyptus leaves for an enormous hung installation, and with the intention of making promises to Country. Developing the work with producer Pippa Bailey and supported by a Carriageworks First Nations Fellowship, Baird added talks and planting, overlaying indigenous cultural protocols with contemporary artforms, weaving plant information with "yarns," in an attempt to reconnect what colonialism has disrupted and separated: "you get the full picture when you go and see the different works," Baird explained (ABC Listen, Awaye!, 2024). The layering and then interweaving of the different knowledges and aesthetics creates a new diagram. In the dance element of *Plant A Promise* – entitled *The Trees Have Voices The Feet Have Ears* – black lines, as if from burned wood, mark the performers' bodies. Baird, in her ABC Listen interview, speaks of the dance having three sections: fire, smoke, and ashes – overlaying destruction, creative process (plants that require fire and smoke to germinate), and protocols – as interdependent and co-reliant parts of a pattern reliant on alternative thinking and practice.

While Baird has worked diagrammatically across a large part of a continent, Louise Ann Wilson has attended to the metaphorical affordances of the details and expanses of particular localities, using a kind of allegorical mapping (through drawing and walking) that overlays, or places intense experiences in the landscape. In that sense, Wilson's practice is site-specific; but it is also scenographic and metaphorical. She and her company attend to the specificities of place, as in *The Gathering/Yr Helfa* (2014), where the annual processes of sheep farming on a hill farm in Cumbria (UK) were woven into the performance; or, rather, the performance was woven into the existing farming, with shepherds and sheep being performers in the piece.

Wilson and her company weave together three distinct strands of "site-specific, subject-specific and people-specific" research, which they deploy as a "socially engaged scenography.... that seeks tangible, therapeutic and transformative real-world outcomes" (Wilson, 2024, unpaginated). Not only do they address the specificities of landscapes, but they take particular features in a landscape as metaphorical zones in which to address, through their performances, intense human experiences, such as "terminal illness and bereavement, infertility and childlessness" (Wilson, 2024, unpaginated).

"Emptiness" and immersion

For *Warnscale* (2015), Louise Ann Wilson drew on her research for *The Gathering/Yr Helfa* during which she had discovered "how shepherds used the word 'empty' to describe a ewe that has not become pregnant before being 'turned up' to the mountain" (Wilson, 2025a, unpaginated). She then combined her research into experiences of childlessness with looking for "an empty room on a mountain.... 'a place.... Traditionally associated with revelation, transition or inspiration'" (Wilson, 2025b, unpaginated). She found the Warnscale Head Bothy (a shelter left open and free of charge for travellers to use) in the Lake District in the North of England and pivoted the ambulatory performance of *Warnscale* around it. The transformatory quality of this work – changing assumptions about and characterizations of landscapes and human bodies – is reflected in a comment by one participant/audience member who compares experiencing *Warnscale* with a "line from a .... song about a landscape being carved by glaciers moving through it – carving deep valleys and enriching the soil with minerals.... the pain of childlessness felt something like that – it had a depth I had never experienced" (quoted in Wilson, 2025a, unpaginated).

Louise Ann Wilson's most recent work is *Becoming Rock* [2023– current]. Like Whall's *Sheep* (see below) a performance without audience, Wilson is photographed in and with a fissured, weathered, and quarry-gouged limestone terrain in response to her multiple surgeries following a diagnosis of pre-invasive breast cancer.

Similarly, Miranda Whall plotted a pattern, but in her case across the uplands of the Cambrian Mountains in Wales for the *Sheep* part of her *Crossed Paths* project (2017–2018). Whall donned a sheepskin and attached cameras to herself (supplemented by GPS and phone). Accompanied by a camera-laden drone, Whall then crawled for some five or so miles, following tracks left by sheep through heather and boggy ground. Having set up the triangulation of her journey by various recording devices, Whall's progress on padded hands and knees was GPS tracked and filmed from her body and from above, leaving a record of her encounter with the meaningful but disordered tracks of the sheep.

Whall appeared to flatten her action, by recording from multiple perspectives, while at the same time she was following the advice of the writer and walker Nan Shepherd and "found myself going 'in' to achieve what Shepherd calls an 'accession of interiority' leaving the outer world of materiality to reach the inner landscape of the spirit" (Whall, 2017, unpaginated); this was a spirit Whall found in bogs, in the spikes of gorse, in streams. At the same time, by adopting the approximate eyeline of a sheep, Whall lost much of the vastness of the vistas, and dropped much of a romanticization of the landscape that often depends on having a distance from it, unlike Wilson who works with a feminine sublime.

While Whall's imitation of a sheep was somewhat cursory – "I think I just put the fleece on as a kind of gesture of transforming my persona, possibly more from the inside than the outside" (Whall, in Rhys Thwaites-Jones's documentary film *Woolly Maggot*, 2018) – her physical commitment to following their tracks was not. As part of a series of actions that involved crawling as badger, "almost otter" and with various small trees (Oak, Birch, and May) strapped to her back, Whall's *Sheep* performances were less about an imitation of the animals or trees she crawled "as" or alongside. Rather, it was a tracing, in the flow of performers and audiences (invited and accidental), of the relations with and around the animals and trees: "going deeply into the living landscape, ecosystems and interspecies dynamics" (Whall, 2017, unpaginated).

Expanded patterns

At an exhibition accompanying Miranda Whall's *Sheep*, in the Oriel Davies Gallery at Newtown, Wales, in 2018, eight poems, eight texts, and eight GPS line drawings were hung symmetrically on the walls of one room. In the centre of this room, eight Perspex containers of organic samples ("turves" taken from the uplands where she had crawled) were symmetrically arranged, by Whall. In an adjacent room, eight videos, one of each of her *Sheep* crawls, were projected onto eight separate screens. The repeated and orderly numerical organization of the exhibition served to highlight the edge of chaos at which so much of the landscape of Whall's crawls operated: uneven and unreliable ground, sudden inundations of rainwater, random arrivals of voracious herbivores, furious winds. Likewise, the close attention to place and to the immediate "other" (the sheep, the heather, the gorse, the streams) was constellated with the performances, and with commissioned poems, an exhibition, a documentary film, a performative site-visit, eight pieces of commissioned music, drawings, maps, panel discussions, and essays.

Other contemporary performative and performance works explicitly using diagrams or diagrammatical thinking include Nolan Oswald Dennis's *No Conciliation is Possible (working diagram)* (2018), at the Goodman Gallery in Johannesburg (South Africa) (see *figure 3.2*). In this work, a diagram of "the meaning and significance of the notion (and dream) of western compensation" for slavery is presented as "a mind map wallpaper in the tradition of the map room" (Goodman Gallery, anonymous, 2018, unpaginated). A mobile step ladder turns Dennis's wall piece into an implied scenography, with the invitation to actively consider the incompleteness of the diagram up close, if not to make some addition to or redrawing of it.

Similar in its scenography are Barbara Matijević and Giuseppe Chico's *I Am 1984 (2008)*. Here Matijević narrates the relations between different

Figure 3.2 Nolan Oswald Dennis, *no conciliation is possible (working diagram)*, 2018–2025. Installation view, Goodman Gallery, Cape Town, 2018. Courtesy: The artist and Goodman Gallery.

persons, events, objects, and actions characteristic of her experiences of and associations with 1984, representing these as parts of a diagram which she draws on a black wall. Expanding performativity to floral allies, the various terraces of Agnes Denes's *Living Pyramid* (2022–2023) in the Sakip Sabanci Museum Garden in Istanbul (Turkey) were planted with the seeds of local plants, placed to face in particular directions according to the specific needs of the plants for sunlight and shade. The geometrical shape of the pyramid was slowly overcome by the tiny patterns of the individual flowers generating in their totality a meta-chaos of tumbling colours and odours. Finally, the geometrical pattern (with its implied reference to ancient belief systems) wholly disintegrated as the parts of the pyramid were shared to members of the community and spread across Istanbul at the conclusion of the exhibition.

While the curating gallery for *Living Pyramid* emphasized the educational/ecological aspects of the work, there is also a subversive element in the disruption of a historical religious-cultural form. This is related to Denes's work since 1969 in making diagrammatical drawings – including *The Human Argument* (1969/2013), displayed in the Schema exhibition (2023) at the Marlborough Gallery in New York – in which she used mathematics (and geometry) as a "non-erratic additional language.... malleable, unemotional" (Denes, 2008, p. 213) by which to engage with a collating of fragmented human knowing.

A reparative use of the diagrammatic form is evident in the use of hoops in *Just Breathe, Okâwîmâwaskiy* (2024–2025), a collaboration between Cree performer Sandra Lamouche and Melanie Kloetzel, an artist with settler ancestry. In this performance in Calgary (Canada/*Kânata*) an audience, initially immersed in a claustrophobic airplane-in-difficulty/climate-crisis scene, is ushered out of the tension, threat and enclosure of the first half of the piece, into a dark and snowy campus landscape. Framed in a grove of trees, Lamouche dances a slow hoop dance and sings; hoop and gesture becoming enwoven (see *figure 3.3*). The separate hoops of this dance each symbolize connectivity and an unending life without beginnings and ends; together they represent an aspiration "that we will all find our place in one great hoop made of many hoops" (Kevin Locke, quoted in Leizens, 2012, unpaginated). As Lamouche dances, she combines the hoops together into a single global form.

These examples do not evidence anything like a universal recourse to diagrammatics, which is far more present in theoretical and critical writing than in performance practice. They do, however, show that performance makers are using materialized and embodied abstractions – with reference to (often alternative or marginalized) systems of thought – to address climate crisis and social malfunction; diagramming the possibility for worldviews that are alternative to one woven into a procession towards disembodied and dematerialized catastrophe.

Figure 3.3 Sandra Lamouche's hoop dance in *Just Breathe, Okâwîmâwaskiy*. Photo: Fifth Wall Media, Tim Nguyen.

The crisis in place-based and immersive theatre: live(d)ness or liveness

The turn to immersion

Immersive performance, increasingly popular in many countries during the 2010s, might appear on the face of things to be the most appropriate form in which to nurture an intense human/more-than-human engagement with an environment. Its incremental growth in popularity and the hyper-realism of performances like those of Punchdrunk, with their audiences free to wander through and experience carefully and up close detailed rooms or outdoor spaces, would seem to have been facilitating a direct, unmediated encounter with materials. While most immersive performances might be anthropocentric in terms of their narratives, much of the form and practices of this theatre-making still appears to be closely in tune with the kinds of immediacy and alliance-making with materials that we are advocating for here.

Complementary to this, after the millennium, in the face of hypermodernity's alienations and intense stresses, many have looked to immersive actions that were not necessarily within theatre but were more or less performative. Some of these have been inspired by the Japanese practice of *shinrin yoku*, sometimes translated as "forest bathing": spending time in a forest – alone or in a small group, autonomous or guided – giving close attention to the sounds, smells, temperatures and agents of a forest while exposing the body to the chemicals, colours and textures of the trees. This practice would seem to have much in common with the kinds of immediacy and autonomy granted to audiences in some immersive performances, but with a more profoundly ecological context, a "being there" with the forest and its many agents. Nevertheless, the primary focus of *shinrin yoku* and its various Western versions (as with immersive theatre) has always tended to be on the consciousness of the human visitor – sometimes framed as "mindfulness" – rather than on an attention to, or active relation with, the agencies of the forest in their own right. Yes, there is attention to the forest, but it is done in the name of human "wellness," in order to improve the forest bather's attention to themselves; it remains a process of separation.

Insidious humanism

Place-based theatre and dance performances that give attention to their audiences' immersion in space comparable to *shinrin yoku* come under similar pressures to orient towards a distinctly human rather than a more-than-human experience. Indeed, UK-based performance group Punchdrunk have resorted to putting their entire audience in masks to stop them focusing on each other rather than on the sited-performances. The more-than-human may be attended to in immersive performances. However, there are strong existing human pressures and aesthetic forms and practices (ones that Baz Kershaw has characterized in "Western theatre" as "a *double* function in the 'recovery of meaning' about a 'nature' that separates humans from their environment" [Kershaw, 2007, p. 305]). These tend to shape a more-than-human contribution either as a metaphor for the human or as an enhancement, focus, or intensification of the human spectatorial experience.

An example of how such ambiguities can assert themselves over the plans of a performance company and how such pressures manifest is evident in a review of the Louise Ann Wilson Company's *The Gathering/Yr Helfa* (2014). Wilson and her collaborators are known for their conscientious and super-sensitive approach to performance making which is always, as they explain on their website, "made after an extended period of immersion in a chosen landscape" (Wilson, 2025b, unpaginated). During their physically, visually, and sensorially demanding and precisely choreographed site-specific performances – in *Warnscale* (2015), *The Gathering/Yr Helfa* (2014), and *Fissure* (2011) – audiences have been immersed in long and often rugged journeys, engaging closely with extensive landscapes in the Lake District, Snowdonia and in the Yorkshire Dales (all UK). Audiences encounter immersive installations – of materials and human performers – that often point to and highlight the terrain and its more-than-human agencies. The underlying themes of many of the company's performances, such as grief, childlessness and infertility, are ones that they often couch in human terms, yet are open to more-than-human relevance.

However, *The Gathering/Yr Helfa* (created for National Theatre Wales) was an exception, for there the thematic focus was to be "based on the annual fertility, reproductive and life-cycles of the ewes" of a particular farm in Snowdonia (Wales), (Louise Ann Wilson website, 2025, unpaginated). Given the nature of agricultural practices and animal husbandry, the life cycles of these ewes are manipulated by humans to maximize their productivity; in-built into the specificity of the site was a certain pull of focus from the sheep towards the human husbandry. The mixture of participants – some 200 sheep, sheep dogs, shepherds, a brass band, children from a local primary school, a small cast of human actors, and an aerial artist – reflected the complexity as well as the ambiguity of the project.

A review of *Gathering/Yr Helfa* by *Guardian* (and latterly *The Stage*) theatre critic Lyn Gardner – a longstanding champion of innovative theatre arts in the UK – captured some of the ambivalence, describing how "the sheep themselves watch from rocky outcrops... They know that this steep place belongs to them and that we are the intruders." Gardner describes a shifting of attention and focus in the piece away from human interventions – "the unaccompanied voice of a boy singing, to Gillian Clarke's poetic script, to the red woollen tramway that runs down the mountain like a scar" – to "what has always been there.... the fall of water, the moss creeping across tumbled stones." Notwithstanding that "always" (betraying a limited human-centred timescale and a false idea of the fixity and stasis of such terrains) the review then concludes by settling for ambiguity: "extraordinary fecundity and renewal, but also bitter barrenness" (Gardner, 2014, unpaginated).

The mechanics of humanist pressures

Performance makers like Louise Ann Wilson are committed to the specificity of the natural terrains they work in and to immersing their audiences in them. They pay close attention to their agencies: "the fall of water, the moss creeping across tumbled stones." They fore-front more-than-human themes and narrative content. Yet, even for them, there are always pressures, aesthetic, cultural, economic and spatial, that work to modulate relationships to the more-than-human agents they seek alliances with; pressures that are already working before a project begins by limiting any address beyond a human framework ("the primary productive purpose of theatre.... [is] the creation of spectatorship" [Kershaw, 2007, p. 306]) and draining any element of immersion of its immanence.

The social and cultural basis of such pressures includes those of theatrical aesthetics and the problem of spectacle. In her book *Immersive Theatres: Intimacy and Immediacy in Contemporary Performance* (2013), Josephine Machon has explored the ambiguities of, and pressures upon, immersive theatre from the point of view of its audience participants. Just as a plunging of "forest-bathers" into a "natural" environment (distant from human dwellings, even if a product of human management) is a far from simple and unalloyed process, Machon argues that, similarly, the immersing of audiences in intense and proximate engagement with a performance's heavily laden sites is not without paradoxes. Immersion can be at the expense of certain senses, leaving audiences susceptible to feeling overwhelmed and diminished by the spectacle. Machon describes the complex interplay of autonomy and manipulation as audiences are inspired or explicitly encouraged "to feel *feelingly*"; how they may find themselves "addressed 'polysensually'" in a "'full-body' inclusion" that demands they "'relinquish distant and reserved experience' and embrace 'mind expanding' or 'mind assailing' perception" (Machon, 2013, p. 36).

Despite the democratic and phenomenological qualities of, or at least ascribed to, these performance experiences, Machon characterizes them differently. She detects the peculiar quality of immersiveness in site-based performance as unconnected to the specificities of places and their agents. She argues that rather than giving attention to the materials of a place or to its atmospheres, the primary function of an immersive performance is to generate affects as products of its sensual reception by its audience. The specificity of "being there" with the matters of a place are dissolved by these performances and transformed into social capital. The primary production during the performances is of takeaway memories of "being there," that can then be subsequently displayed and exchanged in an audience member's social interactions, or simply "owned" as a reflection upon their having been present at a performance.

Rather than the place and its agents being primary, a performance can be made commodified and transferable through a spectator's "embodied memory of the event." Rather than a liveness, proximity and immediacy available and to be taken at face value, the "'live(d)' of immersive performance is a liveness that is frozen in the moment. As it is experienced it is already an affect of something passed". Rather than directly experiencing immersion in a "full sensorium" engagement, an immersive theatre spectator is thus shaped "by the event-ness and popularity.... 'to undergo', to self-consciously save up the 'in the moment' for later consumption and exchange" (p. 44). The immediacy of place and performance is changed from liveness to what Josephine Machon calls its "live(d)ness" (p. 43). Rather than a present moment of engagement, peculiar to that moment, which disappears in the future, as proposed by Peggy Phelan (1993, p. 146), a new kind of commodified "present" is produced that is fixed in its "lasting ephemerality" (Machon, 2013, p. 44) which can then be possessed, deposited or exchanged.

Acquisitions of cultural capital

Although Josephine Machon does not cite Debord and his critique of a "society of the spectacle," basing her analyses on her own empirical engagement with immersive performances and on an assessment of the avowed intentions of immersive performance makers, clearly something similar to the problems that we outlined in Chapter 3 section iv above ("the cancellation of the future") are at work. While the commitment to specificity, to the detail of matters, and to the liveness and proximity of the audience's encounter with place and performance are all often exemplary in these immersive performances, it is in the relations of production and reception that something else occurs: the present becomes inaccessible except as a memory-still-to-come.

Under the terms of such relations, an audience can become quite desperate and frantic to acquire their most intense experience in case they are later found wanting. It is not that the events are not happening or that the materials are not fully present, but that everyone in a performance site, subject to the relations of a society of the spectacle, is part of "a social relationship between people that is mediated by images" (Debord, 1994, p. 12). It becomes necessary for the spectator to become an acquirer-experiencer of images (memories) for later use. Any naive, ingenuous, and "artless" immersiveness – rather than supporting an encounter with the more-than-human agents of a place – is thus, more than most performance forms, subjected to representations. For it is not the representations that a company performs in the moment that is most powerfully active here, but the images that are synthesized and stored up in the same moment by spectators for consumption and exchange later. What began as a movement in theatre to enhance immediacy and intensity, when subjected to spectacular relations over time, has been made archival and a reification machine.

Spatializing a cordoned nature

In *Sunken Lands*, a book on flooded landscapes – terrains shaped both by an ever changing geology and climate, and more recently by events stimulated by human-initiated land management and global warming – the psychogeographically-inclined author Gareth E. Rees writes of a kind of contemporary spatialization in the face of climate change that is both a containment and a warding off. Rees's most vivid example is a visit he pays to Wicken Fen in Cambridgeshire (UK). 2,000 acres of reed and sedge owned and maintained by the heritage charity, the National Trust. It sits within 4,000 square kilometres of now drained (99% as of 2001) former fenland, whose reduction from a previously marshy region to agricultural land, sustained by drainage channels, dykes and drains serviced by automated pumping stations, has been described by the ecologist Ian D. Rotherham as responsible for "the greatest single loss of wildlife habitat in Britain" (2013, p. 22).

Rees recounts his repeated encounters in Wicken Fen with information boards describing brimstone butterflies and birds "such as willow warbler, chiff-chaff and blackcap," none of which ever become visible to him (Rees, 2024, p. 82). Consistent with our description above of a spectacle that pushes, beyond the subjection of things to their representations, for the replacement of things by their images, Rees notes how "the real place was concealed from my senses. Instead of the thing, I was accessing *data about the thing....* as if they were merely bundles of properties and the environment was simply a container in which they existed" (p. 82).

Unlike authors such as Marc Augé in *Non-Places: an introduction to supermodernity* (1995), who have noted similar replacements but mostly restricted themselves to moral censure, Rees is sensitive to ideas of hauntology and psychogeography, allowing him to expand and deepen his cultural observations to ones of embodied sensitivity and spatial insight. Drawing on the work of ecologist and philosopher Timothy Morton, Rees describes how a survival-based strategy developed just after the Ice Age by the first farmers – in which "whatever made the land more productive was prioritized" – has been so widespread for so long. Rees contends that it "leads us to assume it is intrinsic to human behaviour, necessary for our survival and therefore unchangeable" and this, in turn, drives conservationists' attempts to address environmental crisis by ignoring this overall "intrinsic" orthodoxy, and "cordon off bits of land to protect or re-establish wild environments" such as at Wicken Fen (p. 83).

Not only, Rees argues, does this perpetuate "the same pattern of thinking that caused the problem in the first place," it also creates a "false impression of a boundary between us and nature; that there is such a thing as a solid, containable, comprehensible space that we can control" (p. 83). What Rees describes here is something very like the "ideal" form of a theatre or stage space, defined by Peter Brook as "empty"; a spatiality that is inscribed in much of the practice and, indeed, in the very idea of "site" in site-specific theatre and performance (see Smith, Phil, 2025, pp. 52–67). Cordoning off or containerization of a space – warding off extraneous interruptions beyond the control of performance makers, to intensify encounters within it – thus becomes a common "intrinsic" practice in performance, site-specific or not.

When this involves other humans – for example, restricting the movement of homeless people around a performance site – it may trigger unease and protests among audiences and other performance makers. However, where the bounds of a site are invisibly drawn to generate a particular vista, to exclude any "anomalous" agents or to highlight a particular species, the strategy – being "intrinsic" – may be "invisible" to its audiences.

Useful awkwardness

It is against such an apparent "naturalness" of containment and control, and the assumption that places are affordant to human shaping, that a diagrammatical performance sets itself. Indeed, it is the very awkwardness, abstraction, and artificiality of such places' diagrams that can be deployed to "make strange" a space. These diagrammatics are far less "natural" and transparent than the specificities of place fundamental to site-specific and immersive theatre. These diagrams also expose the operations of relations subject to images, undoing (or at least revealing) the unseen machinations of container-space, before beginning an engagement with the place's more-than-human agents.

Gareth E. Rees's conclusion to his consideration of space and culture at Wicken Fen suggests that there is a second and older order of space with which the diagram can also help us. Tellingly, Rees ends with the observation that places, rather than consisting of "solid, containable, comprehensible space" are in fact "mysterious and uncanny, where the human and non-human are embroiled in complex interactions on infinite scales." He cites proponents of dark ecology to challenge the idea that there is any such thing "as 'nature,'" instead suggesting "a teaming multiplicity of entities" among which "[H]umans are objects inside bigger objects.... with smaller objects within us.... Nature is inside us, as much as we are inside it." Having de-naturalized "nature" and proposed that humans are "inextricably enmeshed" in it, Rees advocates for a very different kind of immersion. He proposes pulling down the containing lines of actor/spectator, human/non-human, conserved space/world in order that "we'd each feel every hurtful blow against the biosphere – every oil spill, decimated forest and plume of toxic smoke – as an attack on our own bodies" (Rees, 2024, pp. 83–84).

Transparency and camouflage

In site-specific theatre, there is an idea that, dependent on the nature of a "site" a performance can best attend to its specificities by a light touch (transparency) or by a full-blown embrace (camouflage) of the site's phantasmagoric, spectacular or hyper-performative qualities (Smith, Phil, 2019, pp. 99–106). In either case, if we take seriously the analysis of spectacular relations proposed in the above section "The cancellation of the future", adjusting performance aesthetics to match those of a chosen site itself becomes an interweaving with a quite different set of relations from the ones most site-specific performance makers imagine they are engaged with. By aligning oneself to the apparent agencies and elements of a site in both appearance and essence and honestly addressing the materiality of a place as it presents itself, a performance maker may be doing the very opposite to what they intend. Rather than allowing the place to become the primary performer, a site-specific performance maker may be empowering appearances and relations about their site, whilst distracting from the agents that are "being there."

There is no guarantee that a diagrammatical approach, overlaying a performance to precisely match the appearances and relations of its place, will not unknowingly reify appearances and relations. However, the "making strange" of the diagram, with the awkwardness of its lines, trajectories, empty spaces, and symbols, may serve as a stumbling block to the spectacularized "natural" presence of the site's dominant agents. Thus, both audience and place-agents may be subjected to an ordeal of abstractions in order to become real in performance, and at odds with the relations of a spectacle.

References

ABC Listen. (2024) Interview with Henrietta Baird. Awaye! 7am, 30 November 2024.

Anonymous. (2018) *Goodman Gallery* website. Available at: https://www.goodman-gallery.com/store/shop?ref_id=36635#!. Accessed on: 20.1.25.

Augé, M. (1995) *Non-places: An introduction to supermodernity*. Translated by John Howe. London: Verso.

Bailey, P. (2024) 'Plant A Promise'. Arts on Tour. Available at: https://artsontour.com.au/tours/plant-a-promise/. Accessed on: 19.1.25.

Barreras del Rio, P. and Perrault, J. (1988) *Ana Mendieta: A retrospective*. New York: The Museum of Contemporary Art.

Berger, J. (2009) *Why look at animals*. London: Penguin Books.

Billinghurst, H. (2019) 'Ways of making; producing artworks in the studio in response to experiential walking'. (Doctoral thesis.) Available at: https://pearl.plymouth.ac.uk/foahb-theses-other/321/. Accessed on: 7.1.25.

Blatchford, R. (1894) *Merrie England*. London: Clarion Publications.

Blogg, P. (2025) '2025 Sydney Festival to Honour Country'. National Indigenous Times, 15.1.25. Available at: https://nit.com.au/15-01-2025/15760/2025s-sydney-festival-to-honour-country-through-dance-installation-planting-and-story. Accessed on: 19.1.25.

Bond, A. (2024) 'RHODODENDRONS DESTROYING OUR NATIVE WILDLIFE'. The Woodland Trust. Published 23.5.24. Available online at: https://www.woodlandtrust.org.uk/press-centre/2024/05/rhododendron-destroying-native-wildlife/. Accessed on: 6.1.25.

Bonet, N. (2019) 'Disobeying seagulls: making music with animals. *Com Bufa El Vent*. Edition # 42. Available at https://sonograma.org/2019/04/disobeying-seagulls-making-music-with-animals/. Accessed on: 2.1.25.

British Geological Survey. (2010) 'Is Earthquake Activity Increasing?' Available at: https://www.earthquakes.bgs.ac.uk/research/earthquakeActivity.html. Accessed on: 2.1.25.

Brown, T. (1997) 'Clearances and clearings: Deforestation in Mesolithic/Neolithic Britain', *Oxford Journal of Archaeology*, 16(2), pp. 121–151.

Burrows, D. and O'Sullivan, S. (2019) *Fictioning: The myth-functions of contemporary art and philosophy*. Edinburgh: Edinburgh University Press.

ButCH (Bureau for the Contemporary and Historic). (2024) *Gullfest*. https://37looestreet.org/. Accessed on: 2.1.25.

Clarke, D. and Roberts, A. (1995) *Twilight of the Celtic gods*. London: Blandford/Cassell.

Coverley, M. (2020) *Hauntology: Ghosts of futures past*. Harpenden: Oldcastle Books.

Debord, G. (1994) *The society of the spectacle*. Translated by Donald Nicholson-Smith. New York: Zone Books.

DeGroot School of Business (2024) 'The future of sustainable practices: paving our way to a greener future'. https://www.youtube.com/watch?v=-cndnLfQ8mM. Accessed on: 9/3/2025.

Denes, A. (2008) *The human argument: The writings of Agnes Denes*. Putnam, Connecticut: Spring Publications.

Derrida, J. (1993) *Spectres of Marx*. Translated by Peggy Kamuf. New York & London: Routledge.

DeSilvey, C., Naylor, S. and Sackett, C. (2011) *Anticipatory history*. Axminster: Uniformbooks.

Fisher, M. (2014) *Ghosts of my life*. Alresford: Zero Books.

Fulford, M. (2023) 'Dartmoor belongs to its stories: Old Crockern and protest', *The Gryphon*, 20.2.23. Available at: https://www.thegryphon.co.uk/2023/02/20/dartmoor-belongs-to-its-stories-old-crockern-and-protest/#:~:text=For%20centuries%2C%20Old%20Crockern%2C%20the,impeding%20of%20their%20common%20land. Accessed on: 23.1.25.

Galbraith, J.K. (2001) *The essential Galbraith*. Boston & New York: Houghton Mifflin.

Gardner, L. (2014) 'The Gathering Review'. Guardian Online, 15.9.14. Available at: https://www.theguardian.com/stage/2014/sep/15/the-gathering-review-national-theatre-wales-hafod-y-llan-snowdonia. Accessed on: 7.3.25.

Gershwin, L.-ann. (2013) *Stung!* Chicago & London: University of Chicago Press.

Gibson, J.J. (1966) *The senses considered as perceptual systems*. London: George Allen & Unwin.

Guattari, F. (2000) *The three ecologies*. Translated by Ian Pindar and Paul Sutton. London/New York: Continuum.

Hanson, T. (2022) *Hurricane lizards and plastic squids*. London: Icon Books.

Hays, G.C., Doyle, T.K. and Houghton, J.D.R. (2018) 'A paradigm shift in the trophic importance of jellyfish?, *Trends In Ecology and Evolution*, 33(11), pp. 874–884.

Hurtado, C. and Gallen, S.F. 2024. 'Exploring the impact of deglaciation on fault slip in the Sangre de Cristo Mountains, Colorado, USA'. *Geology*, 13 November 2024. Available at https://pubs.geoscienceworld.org/gsa/geology/article-abstract/doi/10.1130/G52661.1/650062/Exploring-the-impact-of-deglaciation-on-fault-slip?redirectedFrom=fulltext. Accessed on: 2.1.25.

Kershaw, B. (2007) *Theatre ecology*. Cambridge: Cambridge University Press.

Land, N. (2022) *The dark enlightenment*. Baldwin City, Western Australia: Imperium Press.

Leizens, T. (2012) 'Kevin Locke: Elevating the Human Spirit Through Music And Dance'. *Indian Country: Today Media Network*. Available here: https://indiancountrytodaymedianetwork.com/2012/12/16/kevin-locke-elevating-human-spirit-through-music-and-dance-146330. Accessed on: 20.1.25.

Li, Y., Chen, L., Zhou, N., Chen, Y., Ling, Z. and Xiang, P. (2024) 'Microplastics in the human body: A comprehensive review of exposure, distribution, migration mechanisms, and toxicity'. *Science of the Total Environment*, 946. Available at: https://www.sciencedirect.com/science/article/pii/S0048969724043638. Accessed on: 7.1.25.

Lippard, L. (1973) *Six years: the dematerialisation of the art object*. New York: Praeger.
Machon, J. (2013) *Immersive theatres: Intimacy and immediacy in contemporary performance*. London: Bloomsbury.
Madsen, Michael (director) Into Eternity (documentary film), 2010.
Mcphie, J. (2018) 'I knock at the stone's front door', *Parallax*, 24(3), pp. 306–323.
Medecins sans Frontiers (2022) 'Environmental Roadmap'. [Online] Medecins sans Frontiers operation centre Geneva. Available at: https://climatehub.msf.org/wp-content/uploads/2024/06/MSF_OCB_ROADMAP.pdf. Accessed on: 9.3.25.
Met Office. (n.d.) 'UK and global extreme events – Heavy rainfall and floods'. Available at https://www.metoffice.gov.uk/research/climate/understanding-climate/uk-and-global-extreme-events-heavy-rainfall-and-floods. Accessed on: 2.1.25.
Morton, T. (2022) *Spacecraft*. London: Bloomsbury Academic.
Nail, T. (2024) *Matter and motion: A brief history of kinetic materialism*. Edinburgh: Edinburgh University Press.
O'Neill, S. (2022) 'Defining a visual metonym: A hauntological study of polar bear imagery in climate communication', *Transactions of the Institute of British Geographers*, 47(4).
O'Sullivan, S. (2024b) *On theory-fiction and other genres*. Cham: Palgrave Macmillan.
Pascal, B. (1958) *Pascal's pensées*. Translated by William Finlayson Trotter. New York: E. P. Dutton & Co.
Phelan, P. (1993) *Unmarked: The politics of performance*. New York & London: Routledge.
Rees, G.E. (2024) *Sunken lands*. London: Elliott & Thompson.
Rotherham, I.D. (2013) *The lost fens*. Stroud: The History Press.
Smith, P. (2019) *Making site-specific theatre and performance*. London: Red Globe Press.
Smith, P. (2025) 'Three-dimensional metaphor: After site', in: Victoria Hunter and Cathy Turner (eds.) *The Routledge companion to site-specific performance*. New York & London: Routledge, pp. 56–67.
Smithson, R. (1996) 'A provisional theory of non-sites', in: Jack Flam (ed.) *Robert Smithson: The collected writings*. Berkeley & Los Angeles: University of California Press.
Stankus, T. (2008) 'A review and bibliography of honey bee colony collapse disorder: A poorly understood epidemic that clearly threatens the successful pollination of billions of dollars of crops in the USA', *Journal of Agricultural and Food Information*, 9(2), pp. 115–143.
Sydney Festival. (2025) 'Honouring country through dance, installation, planting and story'. Available at: https://www.sydneyfestival.org.au/events/plant-a-promise. Accessed on: 19.1.25.
Thwaites-Jones, R. (2018) *Woolly Maggot*. (film documentary) Available at: https://www.mirandawhall.space/documentary-film/. Accessed on: 19.1.25.
Vousdoukas, M.D., Athanasiou, P., Giardino, A., Mentaschi, L., Stocchino, A., Kopp, R.E., Menéndez, P., Beck, M.W., Ranasinghe, R. and Yeyen, L. (2023) 'Small island developing states under threat by rising seas even in a 1.5°C warming world'. *Nature Sustainability*. 6, pp. 1552–1564.
Wall, T. (2023) 'Thousands march across Dartmoor to demand right to wild camp', *Guardian newspaper*. 21.1.23. Available at: https://www.theguardian.com/environment/2023/jan/21/thousands-march-across-dartmoor-to-demand-right-to-wild-camp. Accessed on: 23.1.25.
Weiss, P. (1966) *The investigation*. London: Calder & Boyers.
Whall, M. (2017) 'CRAWLING – ABOUT'. Miranda Whall -Interdisciplinary Artist. Available at: https://www.mirandawhall.space/crawling-about/. Accessed on: 19. 1.25.

Wilson, L.A. (2024) 'Sites of Transformation (UK)'. YouTube video. Available at: https://www.youtube.com/watch?v=7fNOboe1gSo&t=14s. Accessed on 20.1.25.
Wilson, L.A. (2025a) 'WARNSCALE: A LANDMARK WALK REFLECTING ON IN/FERTILITY AND CHILDLESSNESS'. Louise Ann Wilson Company. Available at: https://www.louiseannwilson.com/work/warnscale. Accessed on: 20.1.25.
Wilson, L.A. (2025b) 'Louise Ann Wilson Company'. Available at: https://www.louiseannwilson.com/about. Accessed on: 7.3.2.
Wright, L. (2023) 'What is Dusking?'. *Tradfolk*. Available at: https://tradfolk.co/art/writing/what-is-dusking/. Accessed on 20.3.2025.
Wynne-Jones, S. (2022) 'Rewilding: An emotional nature'. AREA: Royal Geographical Society, Special Section. Available at: https://rgs-ibg.onlinelibrary.wiley.com/doi/pdfdirect/10.1111/area.12810. Accessed on: 2.1.25.

4 Where did our diagrammatical performances come from?

Being there/becoming there/becoming other

Being there

The practical strategies advocated in this book have been tried, tested, and developed as part of Crab & Bee's slow unpicking of some of our absorbed assumptions and cultural biases with regard to space and place, as two individuals born in twentieth century Western Europe. Residual Cartesian assumptions about body and space are still very present within contemporary language and culture. It has therefore been necessary to have some understanding of their historical and contemporary context before engaging in discussions around a de-colonization and/or re-wilding of "the self".

This is also in part a preparation of theoretical ground, ready for meshing with an historical shift from spectatorial to participatory epistemologies within philosophy. Cultural geographer John Wylie (2007) has described how fifteenth-century innovations around perspective dramatically shifted Western understanding of landscape towards that of an inert, static terrain; measurable, quantifiable, gaugeable in three dimensions, and always subject to the human gaze. This paved the way, two hundred years later, for French philosopher Réné Descartes to further shore up a perceived split between an intelligent but elusive mind, and a body every bit as objectified as its environment (Wylie, 2007, pp. 144–147). Only the mind stood free of this determinism according to Descartes. He argued that even the senses could be deceptive, therefore the reasoning mind was the only reliable locus of human experience. This had profound consequences for ways of thinking about space and place:

> Landscape becomes the accomplice and expression of what is often called a spectatorial epistemology, an approach to knowledge which begins by supposing the following scenario: an external pre-given reality observed and represented from a detached position by an independent perceiving human subject.
>
> (Wylie, 2007, pp. 144–145)

DOI: 10.4324/9781003606673-5

The unpicking of this Cartesian framework began at the beginning of the twentieth century with philosopher Edmund Husserl's criticisms of these ideas as "the natural attitude". His critique was developed by Martin Heidegger in discussions of "*Dasein*" – a compound word meaning "being there." Heidegger argued that human experience, far from that of a thinking mind floating in a void, unfolds as a continuous engagement with the stuff of the world. Heidegger referred to this human interaction as a "being-in-the-world" (Heidegger, 1962, p. 84). Although keen to root lived experience firmly within the world, Heidegger's conceptualization of the world was essentially static and human-centric. His discussions of Dwelling, "tied to 'belonging', 'identity' and 'authenticity'" (Malpas, 2013, p. 2), were deeply problematic, and later used to license German Nazi expansionism.

In the middle decades of the twentieth century, experiential phenomenologist Maurice Merleau-Ponty attempted a new dismantling of Cartesian thought, placing the locus of human experience firmly within the corporeal. He saw bodies as sensory, fleshy conduits through which perception of the world is worked. By pointing to the fact that one hand can touch the other and that we can (mostly) hear, smell, and see other parts of our bodies, in his thesis of reversibility Merleau-Ponty argued that the body is simultaneously both subject and object; thus equally intertwined with mind and world. The Cartesian disembodied gaze no longer seemed viable (Merleau-Ponty, 2014, pp. 81–83). However, as Wylie is quick to emphasize, despite the endeavours of experiential phenomenologists and the turn in this century to vital materialism, the Cartesian framework continues to underpin much Western mainstream media and daily discourse, imbuing scientific and cultural thinking with many assumed dualities, including mind/body, thinking/sensing, subject/object, and inside/outside (Wylie, 2007, p. 147).

Given that our language continues to be littered with such binaries, it can be extremely challenging to dismantle these attitudes within our own discourses and practices. It is partly because of this that we frequently focus on embodiment and sensation in our practical methods, taking on the intuitive, instinctive, dreamed, and fictioned, whilst endeavouring to maintain a critical and interrogative stance.

Becoming there

Shifts away from spectatorial modes of thinking also unfolded throughout the early twentieth century within a range of social sciences. As early as the 1920s, anthropologists such as Bronislaw Malinowski and Margaret Mead had established participatory observation as key to their field research; living amongst the peoples they wanted to learn about and involving themselves in the cultures they studied over sustained periods of time. During the latter half of the century, anthropologists became more and more uncomfortable with the colonial underpinning of their discipline, increasingly rejecting the ethno-centric approach (viewing others through the norms of one's own culture), and more inclined to study all cultures through participatory methodologies.

During the latter half of the twentieth century, revolutionary approaches to psychoanalysis were similarly championed, including those by Jean Oury and Felix Guattari at La Borde, France; an asylum that took a non-hierarchical approach, with participation in many daily activities – from group analysis to washing the dishes – shared by patients and staff. Guattari believed that the challenge for institutional psychiatry "was to learn alongside the patients a new way of relating to the world"; he saw the community as a "mutual thinktank" (Cruz, 2015, unpaginated). Patients were free to move unrestricted throughout the grounds and building.

Fernand Deligny took a similarly radical approach to treating children with non-verbal autism in a rural therapeutic community in Cevennes, France. Encouraging them to wander at will, Deligny believed that by tracing and mapping their walks it was possible to learn about the desires of young people who did not verbally express themselves (Ogilvie, 2015, p. 13). Through these tracings, "distinct patterns began to emerge: certain patterns tended to be repeated from one day to the next." Deligny regarded these as a kind of preverbal language and intuition of place, noting that "some of the wandering lines seem to correspond to the conduits of underground waterways" (Hilton, 2015, unpaginated).

After *les événements*, the student uprising of 1968 in Paris, Guattari developed his ideas around non-hierarchical structures of power with philosopher Gilles Deleuze, publishing *Capitalism and Schizophrenia* in two parts: *Anti-Oedipus* in 1972, and *A Thousand Plateaus* in 1980. In this seminal work, Deleuze and Guattari critiqued Freudian analysis and addressed flows of capital and power, and counter-energies, by introducing concepts such as the "Body without Organs" and the "Line of Flight," while unpicking ideas around human exclusivity and superiority with their discussion of "becoming animal" (1988, pp. 165, 10, 256).

Since the turn of the twenty-first century a move occurred "to redirect attention from the posited meaning towards the material compositions and conduct of representations" (Dewsbury *et al.*, 2002, p. 438), called non-representational theory. Advocates of non-representational theory in human geography such as Nigel Thrift and J. D. Dewsbury argued for increased emphasis on the practices and processes of knowledge production. They proposed the world as a constantly evolving set of relations, too complex to be accurately represented in single outcomes such as a report, book, or map. Instead, they argued for hybrid and experiential practices that emphasize the elusive, relational, and complex elements that make up actions and engagements. Many such experiential tools are now habitually used by archaeologists, ecologists, and geographers as part of their regular methodologies and modes of knowledge sharing; and we have availed ourselves of some of these.

Attention to posthuman, ecological, and decolonialist perspectives continues to identify and confront human-centric, colonial, and capitalist power structures, placing further importance on the rights and agencies of the more-than-human and on indigenous knowledge systems and their "acts of transfer" (Taylor, 2003, p. 2). An emphasis on the temporal and performative, combined with intra-corporeal, material and ecological entanglement, has been underlined by vital materialists, approaching "Lifeworld" as "a meshwork of interwoven lines of growth and movement" (Ingold, 2010, p. 3), advocating "a materialism in which matter is figured as a vitality at work both inside and outside of selves" (Bennett, 2010, p. 62), and placing emphasis on "tentacularity," "multi-species collaboration" and "speculative fabulation" as parts of a process of "staying with the trouble" (Haraway, 2016, p. 71).

Becoming (o)there

Based on these general tendencies within critical theory across a range of fields and, most importantly, on the basis of thousands of hours of field experience, we suggest that the first step to creating diagrammatic performance is to go outside and get amongst things, get a little dirty, get very dirty, entangle with a more-than-human world, pay attention to its forces, find out what is happening out t/here. As Tim Ingold suggests:

> [A] focus on life-processes requires us to attend not to materiality as such but to the fluxes and flows of materials. We are obliged, as Deleuze and Guattari say, to follow these flows, "tracing the paths of form-generation, wherever they may lead"
>
> (2010, p. 3)

To meet such an obligation we deploy tactics from a certain kind of self-conscious walking associated with psychogeography and walking arts. Like Tina Richardson, who created "schizocartography" to study and critique the landscape, we have created a bricolaged methodology, deploying psychogeographic methods (such as the *dérive*) as elements of a toolkit that allows "flexibility in regard to interpretation," while borrowing "from differing theories and practices in order to create a flexible set of instruments." Such a toolkit "can be applied to all stages of the process of analysis, from the physical field work to the critique and research, through to the forms in which the outcomes may be presented" (Richardson, 2017, unpaginated).

Initially – long before we considered making any performances, let alone diagrammatical ones – our intention in our walking practice was, as something similar to Richardson's "instruments," to seek out the spaces of the urban peripheral. Referred to as "edgelands" by Marion Shoard (2002), or "voids" by radical architect Francesco Careri (2002), such spaces are often cited by walking artists and psychogeographers as key to certain kinds of experiences. The material messiness, strange assemblages, and often nonsensical signage of these edgelands disrupt our spatial assumptions and take us to our sensorial edges, connecting us to the under-currents of the human unconscious and the intense social melancholy of abandoned futures.

Friction between extractive human industry and non-human habitats is exposed in these zones; they make for uncomfortable and disturbing encounters with the more-than-human. Now these zones are changing. As industrial infrastructure, city planning, and factory farming methods have expanded to meet the needs of increasing populations, traditional divisions between rural and urban have become less possible to draw. At the same time, ongoing neo-liberal technocratic "austerity" measures have been the drivers of a hollowing out and dereliction of many city centres across the globalized world; changes that intensified after the 2007–2008 banking crisis and were accelerated by a worldwide recession following the 2020–2022 Covid pandemic. Whereas the troubled and theatrical spaces of the edgelands were once found at the literal edges of cities and towns, now these abject spaces are everywhere and more readily available to all.

Space performs

The architect Dorita Hannah tells us that "*space.... performs*" (2011, p. 54). Hannah suggests that scientific and cultural innovations of the twentieth century shifted perceptions of space and time from a "*static spatialisation of time* to the more dynamic *temporalization of space,*" allowing new insights into space as an event, and shifting emphasis from being, to becoming. Despite this – and it is important to see that most of the ideas we celebrate here are parts of an outlying and fractional minoritarian intellectual trend rather than an accomplished cultural change – the current and anachronistic commodity-focused ontologies of anthropocentric, extractive discourse continue to refer to the environment as inert and passive, encouraging those with power to act as if it were so.

The idea that space performs is, maybe, initially surprising and difficult to grasp. Yet if you attend for a while, you will see how space *does* perform, and who and what are the actors there. Shadows shift as the sun moves through the sky and descends below the horizon, and stars begin to glimmer in the evening sky. Wind blows, dust settles, grass grows, and the brambles creep. Tiles blow off roofs, water drips, and ceilings fall in. Objects decay and fall apart, challenging the symbolic order in material terms. Stand in a factory or an office, when work has finished for the day, and listen to the sounds of the building settling down. The longer and closer you attend, the more you will see how "space is an intricate and active player in our lives" (Hannah, 2011, p. 55).

The kind of experience or – better still – encounter that we seek, shifts an act of "*being* there" to one of "*becoming* there", moves us from observer to participant. Key to this is the physical act of placing our bodies within the territory and attending carefully to the agencies at work within the space and upon us. The countless available methods and tactics to help us tune in to the more-than-human-actors and the active, affective agents of a place are all of an embodied kind (Overall, 2021; Hind and Qualmann, 2015; Smith, 2014). During the 2020–2021 COVID-19 lockdowns, for example, there was much discussion of the physical and psychological therapeutic benefits of "wild swimming." Less acknowledged was the opportunity it affords to entangle with the watery world through visceral, embodied immersion. As social anthropologist Sean Heath shows in his research blog about sea swimming,

> What became immediately apparent to me when hopping into the sea clad in a wetsuit for the first time in UK waters was the barrage of sensory experiences, so unlike those I experience in the pool. The cold water stung my feet, the crash of waves breaking against sand and pebble beach produced an auditory cacophony, the reflection of the sun as it danced along the wave tops blinded me through clear goggles, and I could barely see my own hands with the lack of clarity to the water as sand was churned up by the movement of the tides.
>
> (2022, unpaginated)

What is crucial here is that the writer-swimmer does not simply note the light, the crashes of waves, and the churning sand, but has their perception interfered with by them. They are in cahoots with light, waves, and churning sand. These are not simply things to be added to the swimmer's perception; rather they actively interrupt and impose themselves into those perceptions. They are agents doing things; just as much as the swimmer's swimming, the place's crashing and churning and glinting is a doing (and an undoing).

Walking

Our preferred route to "*becoming* there" is through the "simple act" of walking. Not an everyday walking-to-work kind of walking, but a hypersensitized mode that shifts awareness to everything around us that is sharing in the walk. In the first instance we drew from the psychogeographic strategies of the Lettrist International (1952–1957) and the Situationist International (1957–1974), strategies defined in 1955 as "the study of the precise laws and specific effects of the geographical environment, consciously organized or not, on the emotions and behaviour of individuals" (Debord, 1955, p. 8). However, our methods have come to diverge from that definition, in that they are less concerned with the psychology, emotions, and discrete subjectivities of human individuals, leaning rather towards exploring, in an embodied way, the other dynamic and vibrant agents in active terrains.

With "hypersensitised walking" (Smith, 2015, pp. 52–57), the whole sensorium is deployed. As well as careful observation, there is deep listening, tactile investigation (of texture, temperature), sniffing the air (sweet scents/foul odours), and of course exploring taste (there is much to be gained from lifting a smooth pebble from the beach to explore its dusting of salt with the tip of your tongue or tasting blackberries straight from the bramble). Whoever told us that there are only five senses was deluded. Walking draws forth a sense of direction, and a feeling for danger (feel the hairs prickle, the sweat break out). Proprioceptive sensors switch into action as you move your body through a changing landscape; adapting to the terrain with each step on a rocky path and each stoop under overhanging boughs. Walking silently, watchfully, brings forward the rhythm of a route; sometimes straight, flat, dull, or tiring, and on other occasions winding, delightful, and surprising. The sensitized walk can make you aware that some places will lure you in seductively, while some will repel you, push you away. In hypersensitized mode, we question and test the edges of our bodies: where does a sense stop, and the world begin? Is there an in-between? How many agents are there? Can we see everything we need to? Sitting with such affects for a while attunes us to the swirls and forces, currents, unseen tugs and pulls, of a world around us.

Repeated walking in this mode will eventually help to tease out previously unnoticed patterns and codes. Remembered place-times announce themselves; a recurring street-name, an evocative lamppost, and the familiar lay-out of a new terrain steps forward; marking, maybe, the repetition of red post boxes situated at countless crossroads by the village greens, churches, and pubs across rural Britain. Or there are the ruined limekilns at every abandoned river quay in the West Country. Small cavities on the sides of footpaths that were once quarries pockmark the land, often functioning as precursors to nearby mining on a vast, industrial, horrifying scale.

It is worth stating that, while we often think of our walking as our first steps to interdisciplinary creative practice, we are increasingly convinced that walking, and repeatedly returning to re-walk a space, has immense value in itself. Were we only ever to drive to a site, wander around for a while, and then drive off again, the intense insights and understanding of ecological, material, and cultural flows and contexts offered by our ambulatory methods would likely not be available. By walking repeatedly to, across, and around certain terrains, we have tracked the hidden rivers in our city (often now culverted sewers or storm water drains) from source to estuary. We recognize the hyper-stimulation by narcotics and fertilizers of algae and plants around the water treatment plant outflows. We know of the furtive fly tipping in forest valleys, and of rusted scrap cars crammed into the crevices of old quarries. We have seen the orange run-off that leaves Hemerdon tungsten mine and where it enters the Tory Brook. We witnessed the illegal disposal of waste in the reens and ditches of the marshlands around Llaneirwg/St Mellons, on the edge of Cardiff (*Cymru*).

Places can offer up information in the form of ambiences, atmospheres, and other sensory clues to the flows of their vibrant material nature. One winter's day, walking by the Cornish River Seaton, we both remarked on a metallic taste in our mouths. Later, we read of how this river was poisoned at its source by historic copper and lead mining on the edge of Bodmin Moor, the impact of this toxic discharge continuing to affect the river's ecology. The sensual engagement inspired the research. Repeatedly walking a terrain and approaching its key points from new perspectives adds to our ever-evolving networked inventory. There is always information to be gleaned from the most "disreputable" and unscientific of local sources, such as church pamphlets or village noticeboards (not to speak of comments in the local pub); clues to collective trauma or ecological disaster are carved into monumental stones and folded into local myths and place-stories waiting to be further unpacked and critiqued through yet more explorations of a locality in combination with our desk-based reading.

We have found ourselves increasingly less inclined to refer to "space" or "spaces," as the term still holds residual connotations of blankness and emptiness, as if there are empty containers waiting for humans to fill them with something. Despite the argument of theatre director Peter Brook for the "empty space" of a stage awaiting performance by humans (1968, p. 9), no such "empty space" exists. Rather, we speak more of "places" as thready, multiplicitous gatherings of what cultural geographer Doreen Massey referred to as "spatio-temporal events" or "stories so far" (2005, p. 130), and we understand ourselves – as performance makers and walkers – as intrinsically entangling within these spatio-temporal webs, undoing and re-doing them, undone and re-done by them.

Francesco Careri has discussed a state of "hypnosis by walking" which he refers to as "deambulation" and suggests that it is in just such a state that we can "enter into contact with the unconscious part of the territory" (Careri, 2000, p. 82). For the "sensory world of the body awakens us not only to sensation but also to a slower, deeper landscape beneath the surface of everyday awareness, a landscape of feeling, memory, impulse and dream" (Tufnell and Crickmay, 2004, p. 3). There one may experience "a sense of a visitation of non-human agencies (the animals and plants, but also of 'larger forces' at work)" (O'Sullivan, 2024, p. 124).

Certainly, as well as embodied encounters with the flora (nettle-stings, bramble scratches, picking blackberries), fauna (fear of cows and dogs, the irritation of a cloud of gnats, the elation of a starling murmuration), we have become accustomed to unexpected encounters of a "fictioned" kind; entangling with local saints or water wyrms in the Cornish holy wells, or encountering Pan in a children's playground. Such an "imaginative connection" (O'Sullivan, 2024, p.117) may occur when met by a powerful ambience presenting as what we have come to call a "personality of place" (a *genius loci*). Or when a place-story seems to play out in front of us, or a character from such a story suddenly appears to stand forward in solid form. Experiences that can be equally unexpected and unnerving.

There is potential here to become bogged down in an ontological wrangle about where such an encounter is sited: is it subjective or objective? Is it "in our heads," or "out there in the world"? Here, we have found it useful to draw from Simon O'Sullivan's description of art processes that "speak back"

(O'Sullivan, 2016, p. 5), producing images and characters that are equally of the self and yet step forward autonomously from elsewhere. Far from needing to resolve whether such conjurations are the "product of our imaginations" or "actually out there," we have come to view these ambiguities *in themselves* as the first impulses – strands of connectivity or lines of desire – towards some stitching together of the Cartesian split. As such, we see them as useful devices for the production of new subjectivities through creative practice. And it all begins with being there, moving through, becoming with; generating different kinds of knowledge while "on the hoof," reading the territory, feeling out the local stories, joining up dots that will oblige us to some kind of performance that sidesteps representation and mimesis and takes us to the places, stories and characters and puts the places, stories and characters into us.

Making, painting, drawing, walking, mapping

Flexible versions of making

An interdisciplinary approach to performance-making and switching between artistic disciplines can help to bring "something to the surface" (Warzer, 2022, p. 35). Varying processes, changing between media and/or shifting scale – while staying with a theme or enquiry – can help to tease out unnoticed patterns and to draw out previously unrecognized or latent elements of a practice. This strategy can be particularly useful for creatives exploring or thinking in terms of systems, networks, ecologies, assemblages, or constellations, when it renders perceptible the diagrammatic "machine that is almost blind and mute, even though it makes others see and speak" (Deleuze, 1986, p. 34).

Artist academic Johanna Drucker draws a distinction "between visualizations that are representations of information already known," and those "that are knowledge generators capable of creating new information through their use." So, diagrams, in this second expanded sense, "have a dynamic, open-ended relation to what they can provoke," demanding to be "actively engaged and 'worked on'" (Drucker, n.d., p. 65). Artist Steve Baris proposes that such diagrams "prompt active engagement of the viewer," that they allow "viewers to mentally 'move' around [...] abstract space," and as such are performative processes, operating "more like verbs than nouns." Baris also argues that for many artists, the "diagram's very structure of knowledge and perception has become a privileged mode of visualizing the networked life world of the modern era" (Baris, n.d., unpaginated).

So, there is always potential for new findings about a subject, when we change how we diagram it. For example, visual artist Loren Munk splits the different activities of his practice across separate identities and modes of making. As James Kalm, art vlogger, he records "spontaneous gallery and museum visits" across New York City (and occasionally further afield) for his YouTube channel *The Kalm Report*. As painter Loren Munk, he uses his expansive knowledge of the New York Art scene – gathered through his video "fieldwork" repository – to create oil paintings "that portray neighborhoods, movements in art history and philosophical views of esthetics" (Munk, n.d., unpaginated). Paintings on his website are listed under the categories of "Art World Maps," "Artistic Flow Charts," and "Ontology of Art," and the documentarian (not to mention diagrammatical) quality of his painting is represented in titles that include *Bushwick an interim attempt at Documentation* (2012), *Minimalism-Expressionism-YBA* (2014), and *The Subjective Invading the Objective (study 2)* (2017).

While Munk's paintings display highly conceptual information as two-dimensional diagrams, the work also clearly celebrates the haptic materiality of painting; colour pallet and lay-out are carefully considered and rendered in thick oil paint. While the conceptual elements of this work are engaging, witty and thought-provoking, it is very clear that as Munk, his primary intention is to create a good painting. Indeed, this obsession with the substance, construction, and tactual affects of the paintings of other artists is equally demonstrated in *The Kalm Report*; with lingering examination of the sides of stretchers, and close-up shots of staples, nails, blobs, scrapes, and gluey brushwork.

Diagrammatical thinking and drawing

Steve Baris suggests that while many artists are thinking diagrammatically, "the majority of these artworks" may not "look like conventional diagrams." An example of this is the "home made" *Structural Differential* (n.d.) by John Cussans. This assemblage of sieve, red plastic balls, string and paper serves to "explain" twentieth century semanticist Alfred Korzybski's physical chart of the same name and demonstrates how the human nervous system draws abstractions from its environment. Cussans's three-dimensional haptic diagram is a particularly pertinent teaching tool for art students who are struggling to grasp the relevance of theory to their practical studies. It facilitates "demonstrable access to non-verbal, phenomenal and temporal dimensions of living experience and cognitive processes" (Cussans, 2012: online). As well as conveying complex concepts, Cussans's object assemblage is a neat, witty, tactile object in its own right.

So while Munk uses the space of the canvas in a diagrammatic way; to explore, map and expose information that is implicit to his social interaction and embodied movement through the streets, studios and gallery spaces of the New York art scene, John Cussans creates a cool, tactile, object assemblage to interrogate and demonstrate diagrammatic theory to Fine Art student makers. In both cases, however, the "map *is not* the territory" (Korzybski, 1931, p. 750); each exchange is creative, generative, and pedagogical, not substitutional, and "makes something perceptible that was not otherwise" (Warzer, 2022, p. 36).

If we turn this diagramming back to the business of making performances in response to landscape systems – such as ecological milieux, river complexes, path networks, and storywebs – then the drawn line becomes a meeting place for thinking and location. The art historian Norman Bryson tells us how:

> The drawn line in a sense always exists of the present tense, in a time of its own unfolding, the ongoing time of a present that constantly presses forward.... the drawn line presents Becoming. Line gives you the image together with the whole history of its becoming-image.
>
> (Bryson, 2003, p. 150)

In this way, a pencil line traced across a paper surface can correspond to a walked line across the landscape:

> Silvery lines on the green grass were made by my own footsteps through playing fields and pastures. White chalk tracks were eroded by cumulative travellers along ridgeways from Cambridge to Wiltshire, and similarly, feet and hooves etched out the long, red, winding lines of the paths of Devon. Points on the landscape, once only accessed by car, became connected through the multiple lines I drew between them by walking; 'figuring out' on a grand scale.
>
> (Billinghurst, 2019, pp. 106–107)

The idea that performed walking can concurrently be drawing was elegantly and simply exposed by Richard Long in *A Line Made by Walking (1964)*; a straight line made in grass by the artist repeatedly walking it up and down and then photographing the result.

Artist-academic Iain Biggs suggests that we need to return to the origin of the word "drawn," meaning a "transversal" movement. We can be "drawn together into new constellations, despite and with our differences; or drawn into otherness as a result of following threads of intuition, argument, analysis, and so on" (Biggs, 2014, unpaginated). Thus, a web of threads held by people standing in a circle can help those people to draw out the relationships of an ecosystem. An ancient symbol may have a relation to the route of a procession through a performance space, or to a hand gesture made during a ritual. A painted line can refer to a journey, a river, or to a gravitational descent.

The processes of painting, however abstract, with its "grounds", colour fields, sedimentary layers, excavations, edges and textures, have rich potential to be an interrogatory method for exploring "landscape." Such successions were interrogated in depth by Helen for a five-year research project (Billinghurst, 2019). During this time, she made many walks on a single journey across England and then created work in the studio in response to these walks. The final body of work included stacks of field drawings for audiences to thumb through, sequences of oil paintings laid out on shelves, hinged wooden "folding paintings" to manipulate, small assemblages, and long concertinas of intricate folded abstracted "drawn journeys." There were several series of small gouache paper "maps," wall-mounted in grids or sequences. Although very different from each other, the individual sets and series were linked by coded colour, and motifs that ran throughout the spectrum of artworks. The final exhibition *English Diagrams* (2018) constellated this work amongst found objects (wood, bones, water, stones) placed inside a chalk circle on the floor, and in the window recesses of the gallery. At the centre of the space was a painted "medicine cabinet" containing white chalk and deer bones. This box was the focus for a "Cabinet Meeting"; a low-key participatory ritual that opened the show. There was a hopscotch game marked out on the floor at the entrance to the gallery; audiences were invited to play the game, and to pick up the objects and assemblages on the shelves.

One of the most unexpected findings from this research was that halfway through the study, after making many walks, writing reflectively about this journey, and then two years of making work in response in her studio, Helen resumed her journeying to find:

> a shift in the way that I was perceiving the landscape. Themes and motifs from the earliest walks, teased out through blog-writing and then developed in the studio, appeared to 'take off' and take on their own autonomy. The resumed journey now appeared to be saturated with meaning.... As the original aim of the research was to interrogate how walking informs studio activity, to find that the studio-based making conversely informed the activity of walking was surprising. It also showed the potential for 'each of these modes of activity' to become a 'critical tool to consider the other'
>
> (Billinghurst, 2019, pp. 281–282)

A second unexpected, finding of the *English Diagrams* research came later; when the colour-coded "lexicon" Helen had developed to create her body of work in response to the landscape, proved equally available and valuable for Crab & Bee to create performance. So now a tangle of coloured threads or ribbons becomes a way to "mark up" a storied site (see *figure 4.1*), a way to interrogate Doreen Massey's description of place as a "simultaneity of stories-so-far" (Massey, 2005, p. 54), and a way to support the re-telling of these stories through diagrammatic performance.

Place stories, story webs, and yarns

Finding place-stories

As academics, we like to think of ourselves as capable, critical researchers sourcing information from appropriate, reliable, and peer-reviewed sources.

Figure 4.1 *Matter of Britain*, Crab & Bee, 2005. Todmorden Folklore Centre. Photo: Nina Yiu.

However, when it comes to our practice of gathering, teasing out, and re-weaving place-stories, we often depend on "disreputable sources" such as hearsay, local gossip, village noticeboards, and church pamphlets. We have tested and learnt to trust gut feelings and hunches when it comes to detecting underground water sources and buried rivers or rooting out old trade routes along back alleys and cobbled lanes. We are constantly surprised by a body's capacity to pick up tacit information, later to be verified at "respectable sources." Combined with scholarly research, fictioning can be a useful way to construct and test historical theses and to explore ecosophical futures. We encourage you to do the same by deploying "disreputable sources" within your own practice (and in Chapter Seven of this book, we will return to this subject).

An example of an ongoing performance-making that draws equally from a "received" experience and from serious historical scholarship, is the rebirthing into public consciousness of the Crossbones Cemetery as delivered by playwright John "Crow" Constable, Katy "Kaos" Nicholls and members of the community of Southwark. On the night of 23 November 1996, Constable, acting in his "literary or shamanic persona" John Crow, received an extraordinary visitation from what he describes as an "agency" or "muse," named The Goose. "A true muse gives you orders and directions, takes hold of you," says Constable and The Goose did just this: dictating verse to Constable that was to become the *Southwark Mysteries* (2020), taking him for a walk around Southwark at three o'clock in the morning, and leading him to the rusty gates of a patch of rough, industrial ground at Red Cross Way (Constable, 2020).

What is extraordinary about Constable's experience is that at this time (yet unknown to him) excavations for an extension of the Jubilee Line owned by the London Underground were turning up skulls and other bones on this land. They had disinterred an unconsecrated paupers' cemetery marked on maps as "Crossbones" and referred to as "the single woman's churchyard" in John Stow's 1598 *Survey of London*. The women buried in this ground had been working in church-licensed brothels and were referred to as "Southwark Geese":

> These "single women", who worked in Bankside's brothels or "stews" were condemned to be buried in unhallowed ground. Yet for five hundred years in this part of South London their profession was licensed and regulated by the church.
>
> (Crossbones Website, n.d., unpaginated)

The work that subsequently flowered from Constable's encounter with this muse (we might refer to it as a *genius loci* or personality of place) was prolific, multidisciplinary, and rooted firmly within the community of Southwark. "It feels to me that it is an organic, living being," said Constable (2020, unpaginated). From the initial channelled/intuited poems, Constable wrote a mystery play; performed in Southwark Cathedral and Shakespeare's Globe Theatre in 2000 and again in Southwark Cathedral in 2010. After a long campaign, the burial ground was turned over to the community, who have since created a memorial garden and turned its red iron gates into a shrine to the "Outcast Dead." In 2004, Constable and Nicholls formed "Friends of the Crossbones," an "informal network of sex-workers, poets, activists, oddballs and outsiders." Crossbones Vigils for the Outcast are held at the Red Gates on the 23rd of every month and the Dean of Southwark conducts a "special service" – an "Act of Regret, Remembrance and Restoration" – in which the church expresses regret for the treatment of the "Geese" at Crossbones every year around St Mary Magdalene's day.

Yarns and storywebs

Anthropologist Allice Legat's 2008 study of the oral tradition of the Tłı̨chǫ people of North-West Canada tells us of "stories to think with" and refers to the trails across the Tłı̨chǫ's territories in terms of narratives. For Tłı̨chǫ people, walking these trails validate the past and renew the stories; "following their predecessors' footprints." They use the stories to map the spatio-temporal landscape. They teach their children the history of specific sites and how to survive in them, long before the children get to walk these routes or visit these places. A place is considered the living embodiment of its story, and through such story-mapping the young people of the Tłı̨chǫ community learn about a place in advance of visiting it. Important components of that story refer to both human and non-human activity that happens there (Legat, 2010, pp. 37–38).

Legat tells us that Tłı̨chǫ stories are told and re-told, forming an intergenerational mnemonic map for navigation held in the mind's eye. Yet the map is not fixed, it is refreshed when the trails are walked. The Tłı̨chǫ weave their own experiences of the places visited into each re-telling, leaving their own footprints for the future (Legat, 2010, pp. 37–38). Tłı̨chǫ children are taught to walk slowly, and to walk with all their senses alert to their surroundings (including those which are behind them) and to remember key markers as part of a narrative (pp. 39–40). This developing of careful observation (similar to the learning of "the guardian," above) is part of the Tłı̨chǫ's attention to the land – particularly in respect of what was once there and the traces the walkers themselves are leaving behind – as they intimately track changes in their environment, by re-telling their stories as they walk the trails.

The use of mnemonic narratives to chart terrain within historically oral traditions is not unique to the Tłı̨chǫ, and is particularly well documented within Australian indigenous cultures. For example, Miles Holmes and Wanta Jampijinja, in collaboration with Warlpiri aboriginal Elders in the Tanami Desert region of central Australia, describe "dreaming narratives [that] follow a fixed path from place to place across the landscape.... called songlines" (2013, unpaginated). Some of these narrative lines are local, others are part of a network reaching thousands of miles across "ngurra-kurlu" (translated as "from country" or "country within [people]"): "Dreaming narratives encode law for country, that is, the rules and knowledge needed to sustain country and also the rules and ethics of human behavior" (Holmes and Jampijinja, 2013, unpaginated).

For the Warlpiri, "[B]eliefs, knowledge, and social rules are actualized rather than intellectualized" and as such, their ecological knowledge and land stewardship is "deeply encoded in social processes" (Holmes and Jampijinja, 2013, unpaginated). For both the Tłı̨chǫ and the Walpiri, nature and culture are inextricably entangled: a sensual capacity for the "production of forms" (Cixous, 1976, p. 2), showing how the intersection of the "imaginal" with the "corporeal", to create webs of making and meaning that are intuitive, connective and rooted in the terrain, is not only present in the practices of artists (see below).

Journeys

Painter and performance artist Jill Smith has described just such a shift from imaginal to corporeal entanglement in her understanding of landscape that came about with a radical change in her approach to her practice as a then London-based performance-maker. In the 1970s, while working as Jill Bruce in collaboration with her husband the artist Bruce Lacey, Smith was visiting ancient sites in the landscape (Smith, n.d.) and began to have a "feeling the sites weren't separate from each other but linked to a living landscape" (Smith, 2019, p. 24).

While travelling with Lacey in a van full of props between performances at festivals such as the Albion Fairs, Smith realized that the journey itself was becoming an important part of her practice. She began "taking 'gifts' from one site to the next to link them back into a powerful network" (p. 24). These visits with Lacey transformed into unplanned rituals with no human audience, in which they "let the site speak to us, then evolved something we felt the place wanted." During this process, documentation became secondary: "[W]e would do our ceremonies 'for real' and then re-enact them for the camera" (p. 24). The documentary outcomes of this work were shown in two exhibitions at the Acme Gallery in London; *Ancient Forces* (1978) and *Towards the Real Reality* (1981).

Eventually, Smith eschewed the props, costumes, and public performances. She split from Lacey and, for a long time would not sleep in houses, remaining outside or under canvas. She continued to journey to sacred sites, travelling at times by coach, walking the "last miles in pilgrimage," and staying on site "for several days, immersing myself completely in the spirit of the place" (2019, p. 43). Much of Smith's work refers to terrestrial zodiacs; the drawing out of celestial constellations, tracing them on the ground. During the 1980s, she made an epic clockwise circular tour – the "Gypsy Switch"; a rotation with neolithic henge Arbor Low at its hub – around Britain on foot and with horse-drawn wagons:

> So when I was given the diagram of the switch I was extremely interested. At the time I understood it to be a year's journey which Irish travellers had made, moving around a cycle of horse fairs. For Gemini it included Appleby – the great horse fair to which gypsies still travel [from] all over the land
>
> (Smith, 2019, p. 9)

In 2024, Smith was commissioned by Edinburgh exhibition space Fruitmarket to create *Zodiac Journey* in key sites across the Outer Hebrides, documented photographically by Mhairi Law. The work, improvised with handmade costumes, and with personal and found objects, is highly diagrammatic and as concerned with the future as the past:

> I tied the photographs one by one along the body of the Serpent. First, my father near the head, then my elder son, my eldest grand-daughter and finally a blank card representing the hope of generations to come. I held the Serpent with the photos now on it, over the river in gratitude for all those generations and hopes
>
> (Smith, 2024, p. 34)

The turn towards journey in performance has been discussed by David Overend (2012) who notes, amongst others, Kieran Hurley's *Hitch,* Lone Twin's *The Boat Project* and, perhaps most significantly, Melanie Keen and Eileen Daley's *Necessary Journeys* (2005) which is a kind of meta-publication "knitting together" a collection of "diaspora narratives" in which global journeys "from China to Patagonia" were taken and documented by artists working in England "to trace a route map through a series of art projects" (Overend, 2012, p. 5).

A similar example of a woven landscape/journey/culture nexus is demonstrated across four films of performed journeys made by Andrew Kötting in collaboration with his daughter Eden, singer Claudia Barton, artist MacGillivray, writers Iain Sinclair and Alan Moore, photographer Anonymous Bosch, and others. While each film documents the chronological procession of a performance-journey, they are all intercut with visual loops and audial repetitions; stutters to "set thought and language free to create anew" (Hardcastle Jones, 2014, p. 2); a mode of "meandering and halting" with potential to "open up new ways of making sense" (Pereen, 2017, p. 183).

Swandown (2012) charts an odyssey made by Kötting's travelling community of creatives in a swan-shaped pedalo from Hastings, at first by sea and then rivers, to the Olympic construction site in London, eventually traversing the Thames estuary and London lock system. The next film, *By Ourselves* (2015) documents a walked re-enactment of nineteenth-century nature poet John Clare's journey from a mental asylum in Epping to search for his childhood sweetheart in Northampton. Clare (played by Toby Jones) travels ninety miles through the twenty-first-century edgelands of Essex and Cambridgeshire (UK) accompanied at times by the straw bear of Whittlesea. The third film *Edith Walks* (2016) re-traces another historic walk, from Walthamstow Abbey back to Hastings. This walk was made by a Saxon queen, Edith Swanneck, to identify and collect the body parts of her hand-fast-husband King Harold from the battlefield in 1066. Finally, *The Whalebone* Box (2019), has the least linear narrative of all and is full of references to spatio/temporal simultaneity: everything happening now. In this film, a journey to return a box made of whalebone – an "animal battery" – from the South of England to the Isle of Harris in the North of Scotland is spliced with the reveries of Eden Kötting including her repeated dreaming of hunting a whale in a forest.

Each of these performed odysseys contains kernels of the others. There are codes, motifs, characters (historical and living), objects, and poetic references that cross, correlate, and weave across the four films, just as Kötting and his collaborators cross and retrace the footsteps of others and themselves. One of the most repeated motifs across all films is that of a changeable "whiteness" manifesting (perhaps akin to Robert Graves' White Goddess) as landscape muse of Albion in different ways: as swan, as walking woman swathed in white, as "eyeballs of milk; completely white" (Sinclair), as snowy mountains, and (from Herman Melville's *Moby Dick* [1851]) "the whiteness of the white whale."

In all the examples discussed in this section, the material landscape is interpolated with spectral strands of song and story accessible as mnemonic and available, perhaps, on the sensory spectrum of the synaesthete. Common to all these storywebs – these diagrams held in the mind – is that they are created in response to embodied journeying. They share in common their potential to orientate participants and audiences spatially across landscapes, and temporally through deep time. They are all concerned with the un/making of community and human/unhuman relations in times of change and crisis, offering potential to connect to ancestral voices through an altered vision of landscape in which everything happens now, and the living travel alongside the dead.

Legend tripping as eco-performance

What is legend tripping?

Diagrammatical performance is an extended form of performance. Both the diagram itself and any performative action on or in it extends out from the formal moment or designated space of the performance. These also reach back through all the preparatory processes – including the making and sharing of notes, poems, drawings, paintings, photographs, actions, and so on – and then forward through invitations to audiences to visit (or re-visit) the preparatory places. There is never a finished product that *is* the performance.

The activity of "legend tripping" is a variably prankish to serious visiting of haunted places and of places of weird encounters with the purpose of re-enacting and/or provoking a repetition of a legend's events. It can serve as, or be repurposed as, a model for performative preparatory visits for diagrammatical performances, and for deploying a diagram of a performance at every stage of that performance's making.

While it is only recently that academics – anthropologists and sociologists, in particular – have taken note of the phenomena of legend tripping under a collective heading, it is a very longstanding practice. In the *Otia Imperialia* (1214) of Gervase of Tilbury, (Ellis, 2004, pp. 117–118), it is mentioned that knights, in response to a popular legend, would travel by night to the Iron Age earthworks at Wandlebury (Cambridgeshire, UK) to challenge a moonlit spirit-knight in combat.

In its simplest form, legend tripping is the practice of visiting "places associated with legends" (McNeill and Tucker, 2018, p. 8), but more precisely it involves "re-enactments specifically motivated by the desire to test a legend's truth.... ritual performances in which participants seek to presence the very experiences chronicled in the legend" (Kinsella, 2011, p. x). Crucially, these legends – of ghostly hauntings, violent trauma, encounters with monstrous beasts – should, in some sense, be "live," so that the tripper is always vulnerable to becoming a part of the latest iteration of the legend. (This is also an important principle of diagrammatical performance; despite the general coolness and distanced quality of diagrammatical performances, there is always a possibility of the performer becoming what they enact.)

Given this volatility, and the possible vulnerability of a "visitor," the practice has been popular among young people keen to test themselves and their companions in terms of courage and sense of self: "just as a car allows teens to explore places that were previously inaccessible, so do legend-tripping locales provide for experiences generally inaccessible in everyday life" (Kinsella, 2011, p. 30). While this often gives a light-hearted quality to such visits, they can have profound consequences. In a very few cases, host communities have mistaken the playful nature of the tripping for something more sinister and have responded aggressively (so some awareness of social contexts and sensitivities on the part of a tripper is advisable). In most cases, however, the consequences are ones of excitement and intrigue (followed by reflection, and maybe disenchantment or continuing wonderment). Often the visit persists as memories shared by and among a visiting group, but at other times it can spiral out beyond the immediate visitors into dispersed public storytelling, online photos, publications, art projects, and fictions.

Role playing

In "legend tripping," as in "LARPing" (live action role play), trippers often adopt, outwardly or inwardly, certain roles in the legend of the place they are in. They seek to become a part of the story-place, not just as visitors or spectators, but as players or characters in creating anew the story-ecology of a place. While this is often characterized as irresponsible or irreverent engagement with places with serious narratives, the open (perhaps naive) approach to the site, when it invites entanglement and the uncertainties of connection, is very much in tune with a diagrammatical approach to an enchantment of place.

In preparing for a diagrammatical performance, performers can similarly consider how any preparatory visiting of places is already part of their performative mode – that they are already, from the very start of their project, part of the cast of characters – and imagine how they could best place themselves on a diagram of performance from the beginning. They may change that self-locating, they may even find that the self they are locating is changing; whatever arises they will likely need to repeatedly address the nature of their role and relation to each place-story as it is "taking place."

Ostension

Folklorists and sociologists studying legend tripping often now draw on the idea of "ostension"; this refers to the presenting or representing of a legend as something real, contemporary, and having actual presence. This term first appeared as a repurposing of an idea of Umberto Eco's from his work in semiotics. He described a kind of miming during social behaviour, giving the example of waving a pack of cigarettes as a way of offering a cigarette to someone. In an influential 1983 paper, folklorists Agnes Dégh and Andrew Vázsonyi (Dégh and Vázsonyi, pp. 5–34) redeployed this idea from Eco to describe how trippers might "wave" a legend to indicate that there was now a new and real iteration of it.

Popular examples of this in media are often of a morbid kind (animal sacrifices, for example) and mostly these turn out to be hoaxes or fabrications by special interest groups (there are numerous examples of fundamentalist Christian groups using such activities and subsequent media reports to stoke fears of "Satanism"). While we do not generally consider ourselves to be part of Legend Tripping culture, an important part of our own engagements with legends has been encountering the sites (wells, trees, stones) associated with the miraculous deeds or journeys of early medieval and mainly "Celtic Christian" saints (that may be borrowings from earlier belief systems). Beyond the media noise, ostensive actions have included complex and carefully crafted or responsive narratives unfolded over a number of years, such as *The Incunabula Papers* (or *Ong's Hat* mystery) involving elements of role-playing games (RPGs) and live action role-playing (LARPing), or the two series of the TV documentary *Hellier* (both 2019) in which a group of paranormal investigators visit the sites of possible encounters with hobgoblins (that may be aliens).

The crucially distinctive quality of legend tripping, whether the visits and ostensions are frivolous or attentive and crafted, is that it commits any visitor to a certain vulnerability to the place-story. It assumes that the story-place is active and agentive, capable even of transforming the persons or circumstances of visitors by its own means. Michael Kinsella describes these places as "distinct locations that, for whatever reason, are perceived as being out of the ordinary.... the places – old graveyards, bridges, caves, or deserted houses – don't usually fit in with the surrounding, familiar environment" (2011, p. 11); these places actively resist the uniformity and conformism of their surrounds (resonating, perhaps, with the motivations of the many psychogeographers that continue to drift towards edgeland spaces). In terms of their attitudes to place, legend trippers are less likely to regard the legend-space as an inanimate backdrop to leisure activities or the commodified vistas and ambiences of much tourism and hiking. Instead, legend tripping is much closer to pilgrimage in terms of an anticipated intensity and entanglement with place, though usually lacking any fixed ideology, whether that is a religious belief or the expectation of self-transformation that is now a part of contemporary secular pilgrimage.

In many ways the "to the side" story-places of legend tripping, as described by Kinsella, are similar to ancient "privileged points" as defined by the philosopher of technology Gilbert Simondon as ancient nodes, structured according to a "mode that is prior to the segregation of subject and object.... as if all of man's power to act and all the capacity of the world to influence man were concentrated in these places and these moments" (Simondon, 2011, p. 412). Where legend tripping advances on Simondon's description is where the trippers identify, by their attention to these places that refuse to "fit in," certain contemporary and live "privileged points" of much more recent construction or telling; places where a contemporary breakdown of subject and object has occurred (thanks to deep fears, unconventional behaviours or weird rituals) in the margins of what is accepted or acceptable.

The supernatural?

While the place-stories of legend tripping are not exclusively associated with supernatural events, they often are. They can also be places of crime, unconventional belief systems, or former social experiments. This can be useful to anyone preparing a diagrammatical performance, as very often we have found that the ecological and material-historical narratives of a place are told in "supernatural" forms: an angry mermaid destroying a port at the mouth of a polluted river, or the intuited presence of "knockers" and hobgoblins where the ground is excavated for metals.

When preparing a performance it is generally a good idea to get away from how the supernatural is mostly portrayed in popular media: as a random, often violent, and usually disturbing intrusion into everyday life from some "other" dimension or unreality. In contrast to these daunting fictional tropes, ethnographical studies among older women in Manchester (UK) in the 1980s by Gillian Bennett (1999, pp. 3, 39–75) make a distinction between modern legends and folklore (which largely follow the sinister patterns of popular media) and the "personal narratives" or "memorates" of supernatural experiences of participants in which their encounters are often comforting, domestic, sustaining of family connections, and friendly. Most of the supernatural characters in these stories – whether deceased family members or more exotic "guardian angel-type" figures – are benevolent and helpful.

Basing his propositions on the findings from Bennett's work, folklorist Timothy Lloyd suggests a theory of the supernatural, which is "more than just the eerie or the mysterious.... I suggest we reconceive the supernatural as the realm of the fundamental patterns and rhythms – of time, space, growth, and decline." Lloyd associates this with purpose and meaning, in a somewhat cyclical and equilibrated manner, but then he expands his proposal further: "Seen this way, the supernatural is not a separate, exotic, largely closed realm; rather it is connected to, and underlies, all life... it is supernatural in the literal sense of being the largest pattern of the natural" (1995, p. 60). It is in a similar manner that we are proposing here not only the use of place-stories that are generally understood as supernatural and chimerical, but also the experiences to be had there, as both giving access to patterns of natural connectivity.

Through Lloyd's characterization of the fanciful and enchanted narratives of often troubled places, we can bend back to our ecological and site-specific intentions. Rather than detaching a performative visitor from reality, the stories and experiences of these places are access points to their connections – whether conceptually, biologically, or geographically – to "a bigger picture," to a web of agents. In other words, although the members of a cast of characters in any place might be told as, say, "demons" or "dragons," that does not commit a performative visitor to a belief in Hell or an alternative zoology in order to stay open to what those characters might mean or assert. Instead, there is no need here for either a clear distinction between the everyday and the supernatural in a single liminal space. Rather, we can appreciate and value the ambiguity itself connecting the real and the unreal, the everyday and the miraculous. That feeling we can have of "*did that really happen, or did I imagine it*?" need not be collapsed into a single answer. If we recognize such an ambiguity as the prime zone of connectivity – as we do specifically in relation to the formation of haecceities (see Chapter 5, ii, below) – it can be a means to accessing the "largest pattern of the natural" (Lloyd, 1995, p. 60). The reality status of the place-story is thus less important than the access it gives us to ecologies in crisis.

The diagram's utility

The usefulness of a diagram here – even in its simplest form – may be self-evident. If not, using this book should increasingly bring its utility to light. For a diagram can allow a performance maker to work with the shape of the often elusive, fragile, or daunting connections turned up by embodied visits to troubled places, without having to surrender either rigorous critical scepticism or a sense of enchantment. Rather than framing what the performance maker learns from a place within, outside, or against a belief system, as real or unreal, as material or imaginary, the abstraction of a diagram allows them to place it as part of a mentally pictured structure. The material nature of diagrams – things that can be drawn or assembled – enables the performance maker to literally place what they have experienced or learned in, on or within a material frame in performance or art making. At the same time, the ambiguity of the diagram – as simultaneously abstract and material – allows the elusive thing learned or experienced to retain its ephemerality, without disappearing.

An example of this arose during Crab & Bee's working in Llaneirwg/St Mellons, a suburb of Cardiff (Wales). Here, with community members of Common Wealth Theatre, we uncovered a path, blocked by two steel gates that now divide the modern St Mellons estate from the larger houses at the centre of the old village. The path begins at the site of "Llaneirwg" ("enclosure of Eirwg"), a former earthen structure that formerly stood on the present site of the hilltop church of St Mellons. The path crosses the old Roman road between Cardiff and London and then runs down though the modern estate over the Roman sea defences to the waters of the *Môr Hafren*/Bristol Channel, where Iron Age trade and travellers were connected by these waters to what is now "England."

Eirwg was a community chief during the second century (CE), in legend a great-great-grandson of Bran the Blessed/*Brân Fendigaidd* (literally Blessed Crow), King of Gwent. In a diagrammatical performance (*Nets, Webs & Carrier Bag*, Royal Conservatoire of Scotland, 2021) we told the story of this path. We illustrated the path's connections to Somerset and to London, where Bran's voluble head was installed in The Tower of London, illustrating/performing this journey by carrying a crow painting of Helen's along a thick black ribbon to "London." As we performed this action, we refuted the Roman accusation that such veneration and supernatural consultation of ancestors' heads was a brutal collection of trophies from enemies' bodies. The diagram enabled us to retain the materiality of the path and its connections, while inserting its place-story into a web of crow-based narratives (of flight and flows), the "magic" narrative of Bran's decapitated head woven into the physical class-based separation of St. Mellon's estate from its hill (the headland of Eirwig).

One role of a diagram is thus to enmesh any human architecture or aesthetic fabrication with existing "patterns of the natural." It retains, across both its abstract and material mappings, patterns of multiple connections of the natural and the revenant of the physical makings of landscapes by humans across thousands of years. Such a multi-layered pattern sidesteps overarching questions of belief or faith in favour of the investigation of places, place-stories, and experiences. No one, for example, has ever questioned whether Bran was historical, nor if his head really spoke, nor asked us whether or not we believe it did. According to this method, the uncovering of the details of a hoax or a fiction (the story behind the story) is every bit as enchanting as the mystery of "unnatural" means, while a biological explanation is every bit as theatrical as tall tales about floating monks whose staffs sprout into oak trees.

References

Baris, S. (n.d.) 'About Diagrams (About Diagrammatical Thinking)'. *Steve Baris*, website. Available online at: https://www.stevenbaris.com/about-diagrams. Accessed on: 2.3.25.

Bennett, G. (1999) *Alas Poor Ghost: Traditions of belief in story and discourse*. Logan: Utah State University Press.
Bennett, J. (2010) *Vibrant matter*. Durham & London: Duke University Press.
Biggs, I. (2014) 'Drawing and place: a provocation'. Available online at: https://www.iainbiggs.co.uk/2014/01/drawing-and-place-a-provocation/. Accessed on: 24.3.15.
Billinghurst, H. (2019) 'Ways of making; producing artworks in the studio in response to experiential walking'. (Doctoral thesis.) Available at: https://pearl.plymouth.ac.uk/foahb-theses-other/321/. Accessed on: 7.1.25.
Brook, P. (1968) *The empty space*. New York: Atheneum.
Bryson, N. (2003) 'A walk for walk's sake', in Avis Newman and Catherine de Zegher (eds.) *The stage of drawing, gesture and act*. London: Tate Publishing.
Careri, F. (2000) *Walkscapes*. Barcelona: Editorial Gustavo Gili.
Cixous, H. (1976) 'The laugh of the Medusa'. Translated by Keith Cohen and Paula Cohen. *Signs*, Volume 1, Number 4. Chicago: The University of Chicago Press. Available online at: http://www.jstor.org/stable/3173239. Accessed on: 9.1.25.
Constable, John 'Crow'. (2020) 'Goose And Crow: Urban Magic'. Breaking Convention. Available online at: https://www.youtube.com/watch?v=85YPGPA0UwM&t=664s. Accessed on: 8.3.25.
Crossbones Website. (n.d.) Available online at: https://crossbones.org.uk/history/. Accessed on: 7.3.25.
Cruz, C. (2015) 'Notes Towards a New Language: On la Borde'. *Poetry Foundation*. Available at: https://www.poetryfoundation.org/featured-blogger/72464/notes-toward-a-new-language-on-la-borde-. Accessed on: 13.1.5.
Cussans, J. (2012) 'There Is No Diagram. A Polemical Demonstration of Korzybski's Structural Differential'. *Banner Repeater*. Available online at: https://www.bannerrepeater.org/recorded-talks-and-lectures. Accessed on: 3.3.25.
Debord, G. (1955) 'Introduction to a critique of urban geography', in Ken Knabb (ed.) *Situationist international anthology*. Berkeley: Bureau Of Public Secrets.
Dégh, A. and Vázsonyi, A. (1983) 'Does the word "dog" bite? Ostensive action: A means of legend telling'. *Journal of Folklore Research*. 20(1), pp. 5–34.
Deleuze, G. (1986) *Foucault*. Translated by Sean Hand. Minneapolis: University of Minnesota Press.
Deleuze, G. and Guattari, F. (1988) *A thousand plateaus*. Translated by Brian Massumi. London: Continuum.
Dewsbury, J.D., Harrison, P., Rose, M. and Wylie, J. (2002) 'Enacting geographies', *Geoforum*, 33, pp. 437–440.
Drucker, J. (n.d.) *Graphesis: Visual forms of knowledge production*. Harvard University Press. Available online at: https://monoskop.org/images/2/2a/Drucker_Johanna_Graphesis_Visual_Forms_of_Knowledge_Production.pdf. Accessed on: 2.3.25.
Ellis, B. (2004) *Lucifer ascending: The occult in folklore and popular culture*. Lexington: University Press of Kentucky.
Hannah, D. (2011) 'Event-space: Performance space and spatial performativity', in Jonathan Pitches and Sita Popat (eds.) *Performance perspectives: A critical introduction*. Basingstoke: Palgrave Macmillan, pp. 54–62.
Haraway, D. (2016) *Staying with the trouble: Making kin in the chthulucene*. Durham & London: Duke University Press.
Hardcastle Jones, K. (2014) 'Deleuze's "stuttering": Decomposition, deterritorialization, and pushing language to its limit', *Kelly Hardcastle Jones: What I'm Making*. Available online at: https://kellyhjones.wordpress.com/wp-content/uploads/2014/03/deleuzes-stuttering.pdf. Accessed on: 13.1.25.
Heath, S. (2022) 'Reflections on Covid: Sensing the environment in outdoor swimming'. *Centre For Spatial, Environmental & Cultural Politics*. Available online at: https://blogs.brighton.ac.uk/secp/2022/01/21/work-in-progress-reflections-on-covid-sensing-the-environment-in-outdoor-swimming/. Accessed on: 7.3.25.

Heidegger, M. (1962) *Being and time*. Translated by John Macquarrie and Edward Robinson. Oxford: Blackwell Publishers Ltd.

Hilton, L. (2015) 'Mapping the Wander Lines: the quiet revelations of Fernand Deligny'. *Los Angeles review of books*. Available online at: https://lareviewofbooks.org/article/mapping-the-wander-lines-the-quiet-revelations-of-fernand-deligny/. Accessed on 14.3.25.

Hind, C. and Qualmann, C. (2015) *Ways to wander*. Axminster: Triarchy Press.

Holmes, M.C.C. and Jampijinja, W. (Stephen Patrick). 2013. 'Law for country: The structure of Warlpiri ecological knowledge and its application to natural resource management and ecosystem stewardship'. *Ecology and Society* 18(3), 19. Available online at: http://dx.doi.org/10.5751/ES-05537-180319. Accessed on 8.1.25.

Ingold, T. (2010) 'Bringing Things To Life: Creative Entanglements In A World Of Materials'. *Realities*. Economic & Social Research Council Working Papers. July 2010.

Kinsella, M. (2011) *Legend-tripping online*. Jackson: University Press of Mississippi.

Korzybski, A. (1931) 'A Non-Aristotelian System and Its Necessity for Rigour in Mathematics and Physics'. Paper presented to the American Mathematical Society, New Orleans. Available online at: http://esgs.free.fr/uk/art/sands-sup3.pdf Accessed on: 4.3.25.

Legat, A. (2010). 'Walking stories: Leaving footprints', in Tim Ingold and Jo Lee Vergunst (eds.) *Ways of walking: Ethnography and practice on foot*. Farnham: Ashgate Publishing Ltd.

Lloyd, T.C. (1995) 'Folklore, foodways, and the supernatural', in Barbara Walter (ed.) *Out of the ordinary*. Logan: Utah State University Press, pp. 59–74.

Malpas, J. (2013) 'Rethinking dwelling and the question of place' [online] https://www.academia.edu/4990079/Rethinking_Dwelling_Heidegger_and_the_Question_of_Place. Accessed on: 25/10/2024.

Massey, D. (2005) *For space*. London: Sage Publications.

McNeill, L.S. and Tucker, E. (2018) *Legend tripping: A contemporary legend casebook*. Denver: Utah State University Press.

Merleau-Ponty, M. (2014) *A phenomenology of perception*. Translated by Donald A. Landes. Oxon: Routledge.

Munk, L. (n.d.) 'Loren Munk/James Kalm'. Available online at: https://www.lorenmunk.com/ Accessed on: 5.3.25.

O'Sullivan, S. (2016) 'On The Diagram (and a Practice of Diagrammatics)'. In *Situational Diagram*, edited by Karin Schneider & Begum Yasar. New York: Dominique Lévy.

O'Sullivan, S. (2024) *From magic and myth-work to care and repair*. London: Goldsmiths Press.

Ogilvie, B. (2015) 'In between the lines'. In *The Arachnean and other texts*. By Fernand Deligny. Translated by Drew S. Burke and Catherine Porter. Minneapolis: Univocal Publishing.

Overall, S. (2021) *Walk write (Repeat)*. Charmouth: Triarchy Press.

Overend, D. (2012) *Making Routes; a Study Room Guide on Journeys in Live Art*. Live Art Development Agency. Available online at: https://www.thisisliveart.co.uk/wp-content/uploads/uploads/documents/MakingRoutes_Final.pdf. Accessed on: 17.3.2025.

Pereen, E. (2017) 'Stutter', in Mercedes Bunz, Birgit Mara Kaiser and Kathrin Thiele (eds.) *Symptoms of the planetary condition: A critical vocabulary*. Luneburg: Meson Press, pp. 179–184.

Richardson, T. (2017) 'Assembling the assemblage: developing Schizocartography in support of an Urban Semiology'. *Humanities*. Special Issue: 'Spatial bricolage: methodological eclecticism and the poetics of "making do"'. Available online at: https://www.mdpi.com/2076-0787/6/3/47. Accessed on: 28.10.24.

Shoard, M. (2002) 'Edgelands', in Jennifer Jenkins (ed.) *Remaking the landscape*. London: Profile Books.

Simondon, G. (2011) *On the mode of existence of technical objects* [extract]. Translated by Ninian Mellamphy. Edinburgh: Edinburgh University Press.

Smith, J. (n.d.) 'Jill Smith (aka Jill Bruce) on Bruce Lacey, the Albion Fairs and sacred sites'. Interview with Richard White, artcornwall.org. Available online at: https://www.artcornwall.org/interviews/Jill_Smith_aka_Jill_Bruce2.htm. Accessed on: 10.1.25.

Smith, J. (2019) *The gypsy switch and other ritual journeys*. London: Antenna Publications.

Smith, J. (2024) *Zodiac journey*. Edinburgh: Fruitmarket.

Smith, P. (2014) *On walking*. Axminster: Triarchy Press.

Smith, P. (2015) *Walking's new movement*. Axminster: Triarchy Press.

Taylor, D. (2003) *The archive and the repertoire: Performing cultural memory in the Americas*. Durham & London: Duke University Press.

Tufnell, M. and Crickmay, C. (2004) *A widening field: Journeys in body and imagination*. Alton: Dance Books Ltd.

Warzer, B. (2022) 'History put in touch with itself: Diagram and re-enactment', *Performance Research*, 7(8), pp. 35–41.

Wylie, J. (2007) *Landscape*. New York & London: Routledge.

5 Our vision for performing space

Apocalypticism and a "Planet B"

Morbid narratives

Despite the accelerating effects of climate change, we have resisted both apocalyptic narratives and assumptions about a "tech fix" for global warming, even when "[i]t sometimes seems that [the] story is approaching its end" (Le Guin, 1986, p. 168). Such "end times" approaches, of both overwhelming threat and surprise redemption for a chosen few, seem to us to be wound into, and generative of, what Ursula Le Guin calls the "Story of the Ascent of Man the Hero"; "the killer story" which is "a heroic undertaking, Herculean, Promethean, conceived as triumph, hence ultimately as tragedy" (Le Guin, pp. 167, 168, and 170).

With both of us living within a few hundred metres of defunct nuclear reactors rusting in the hulks of decommissioned submarines in a naval dockyard (with the weekly Monday morning 11.30 am testing of the nuclear accident sirens), perhaps we are especially resistant to any indulgence in "end times" narratives. Our proximity to the decrepit reactors may also explain why we value, respect, and at times celebrate decay and subsidence – as well as blossoming and rooting – as efficacious if daunting more-than-human agencies.

Kathryn Yusoff in *A Billion Black Anthropocenes or None* (2018) has pointed out how global majority communities have repeatedly, over the last few hundred years, faced existential threats to their lifeworld that have included invasion, colonization, enslavement, extraction, and transportation. Only now, however, that a more affluent (mostly white) minority community faces a similar threat from climate change has it become a matter for global concern. Yet that concern is well-founded. The reaching of a certain dead-end of neo-liberal thinking and politics has left even many industrial and post-industrial societies stripped of adequate public services capable of addressing general crises such as flooding, wildfires, and pandemics. Meanwhile, economic instability and periods of diminishing returns on investment have been countered by governments and corporations attempting to grow monetary economies based on information technologies powered by

DOI: 10.4324/9781003606673-6

extractive industries and funded by massive borrowing against the future; which, in turn, compound problems of global warming. Not to speak of the cultural confusion generated by new levels of disinformation, relativism, and unreality.

"Planet B"

It can seem that humans are collectively stumped and effectively paralyzed by the enormous complexity – "And You May Find Yourself Living in an Age of Mass Extinction" (Morton, 2021, p. 1) – of our situation. It is important, then, to acknowledge the pressure and overwhelming nature of this. Sorrow, anger, and a sense of irony are as much a part of making diagrammatic performance as jouissance, sensuality, and a celebration that we are still here, still connecting: "[G]rief is a path to understanding entangled shared living and dying; humans must grieve *with*, because we are in and of this fabric of undoing" (Haraway, 2016, p. 39). Diagrammatic performances offer ways into a "string figure game" (p. 57) that changes patterns and gathers different stories for our "carrier bag/belly/box/house/medicine bundle" (Le Guin, 1986, p. 169), while engaging with "the trash of the Anthropocene" for making "still possible pasts, presents and futures" (Haraway, 2016, p. 57).

There is a well-worn slogan in climate change politics: "there is no Planet B." The vividness of the threat – do something now or the Earth vanishes! – mirrors the extreme nature of the refusal to adequately notice, let alone recognize or respond to, the crisis. Despite our experiences rummaging about in post-industrial sites, we are not always with terrain that remains blighted, where no birds sing and where the water supports little in the way of life. Sometimes, we find places where plants and animals have found a way to thrive. In formerly ruined landscapes, we have come across odd combinations of trees and "garden escapes" rooting in old diggings that support renewed food chains. When we come across a cow's carcass there, stripped to its bones, it is a sign of the return of a healthy population of other animals. Then our assumption is that a catastrophic Earth, a Planet B, could be similarly uneven; abject but not necessarily lifeless. Indeed, that some more-than-human entities might be even more vigorously active than before, and that there would be a shift in the hierarchies of predation. If humans are still around to see any of this, they would almost certainly no longer consider themselves as the stewards of nature; and have to live more of the time on more-than-human terms.

Given the loopy nature of historical time – with futures begun long ago, and previous generations and cultures returning retrospectively – it makes sense to us to at least partly live on Planet B now. To live – and make performances – as if some kind of Planet B is at least likely, if not in part already here, rather than assume that life will largely go on as it is now. Or simply and abruptly cease. In which case it makes sense to make performances with the waste products and leftovers of extraction and abject industries, rather than rely on a future of continuing production and advancing technologies. Or relinquish all responsibility in the shadow of non-existence.

This is not a kind of "green" accelerationism; we are not cheerleaders for "natural disasters" and the extinction of species. We are not suggesting that, rather than protest the tempo of capitalist depredation of the environment, we should embrace it and speed it up in order to bring about its downfall and take our chances in the remaining wasteland. As Benjamin Noys argues in *Malign Velocities: Accelerationism and Capitalism*, when "history advances by the bad side [it] does not mean we should celebrate the 'bad side', but rather recognize this is the ground on which we struggle" (Noys, 2014, p. 9).

Unpredictable

A problem for representation making – including performance – lies not just in the likely severity of an approaching climate catastrophe, but in its tempo. On *Earthwalking*, a two day ambulatory event in 2015 along the Dorset and Devon coast (UK) organized by artist Anne-Marie Culhane (Fremantle, 2015, unpaginated), climate change scientist Tim Lenton described how the dynamics of climate change might not conform to a gradual intensification of extreme weather, flooding and temperature rise, during which human-made representations of the crisis would intensify and become more impressive, widespread and effective. Instead – in terms of climate as a dynamical system – there might be a gradual and barely perceived approach to the lip around a "basin of attraction," during which extreme effects would be sporadic and isolated, followed by a sudden plunge into the basin, when everything would become subject to a sudden and widespread transformation of circumstances. So, rather than a gradual linear progression of problems, which performance makers might calmly respond to while building an argument for drastic action, the effects of climate change may arrive far too violently and suddenly to allow any such thing.

Sudden changes of rhythm, including those of crisis and catastrophe, are well within the capacity of existing drama-making. Yet the tempo of future events may well preclude us from producing any meaningful work in response. While an "end times" narrative might seem to fit well with such an idea of the abrupt escalation of extreme events, its baggage of punishment, purgatory, redemption, and salvation, and a utopian strand that requires the walling-off of righteous communities, threatens us with various delusions. The tendency of such narratives so far has been to end with a comforting way out, a sprig of hope for a handful of survivors that ignores both the agency of the "flood" of catastrophe and downplays costs to "the losers."

Rather than choose annihilatory doom or unrealistic hope, in our diagrammatical performances we are seeking a "flattened" approach that assumes no detailed future or solutions (or amelioration) but instead presents symptoms and responses across a plane of possible circumstances. Most importantly, though, we are presenting sets of relations and alliances with materials and entities as practical future-in-the-now possibilities.

Haecceities as alternatives to self-expression

Tincombe

One autumn Sunday afternoon, we (Crab & Bee) are walking in the hinterland of the small town of Saltash on the Eastern edge of Cornwall (UK). As often happens we veer off the familiar routes, meander along by a stream and then down the gentle hillside of a park. We find ourselves walking around a set of jagged railings and then peering down through this fence into a small former quarry. At the bottom of the hill, we are able to step into the hole in the hillside to admire its cliff-like walls. It is a small void, maybe thirty metres across. Helen, still on the periphery, watches as Phil walks to the centre of the quarry and – to test the acoustics – claps his hands together loudly twice in the still air. Phil's action reminds Helen of a lord summoning a minion in a movie, so when a few leaves stir unexpectedly at his feet, she laughs and says "look what you have called up!" With that, all the leaves on the floor of the quarry shift sideways and then, gathering together, they flutter into the air, forming a rotating column of leaves that travels across the quarry from Phil towards Helen and – just as unexpectedly – drops to the ground at her feet.

Haecceity

Perhaps it was the vibration of Phil's clapping that caused the air to stir up the leaves; we will never know, and identifying an exact cause is not particularly important here. More interesting is that this encounter took both of us beyond routine experience, disrupting our usual sense of cause and effect; of us, the humans, and this quarry, of how things connect. But just what did we encounter in the quarry that day? French philosophers Gilles Deleuze and Felix Guattari would refer to such an event as a "haecceity".

"Haecceity" can be translated literally as "thisness," but Deleuze and Guattari used the term more precisely to describe "a mode of individuation very different from that of a subject, thing or substance" (1988, p. 287). A haecceity is a drawing together of forces or elements – a bundle of material or bodily affects – over a duration. Deleuze and Guattari state that a person – or a life – is a haecceity; an assemblage-event of biochemical fusion, fungus, ancestral DNA, dreams, friends, ideas, and so on; "a set of speeds and slownesses between unformed particles" (p. 289). These gatherings of connection often manifest most intensely at specific times, like "a cloud of locusts carried in the wind at five in the evening, a werewolf at full moon, a vampire who goes out at night" (p. 289). The timing of the haecceity is as important as the space and mass of the material forces that gather as a "thisness."

So, on that autumn Sunday afternoon in Tincombe Park, near Saltash, when Phil clapped his hands the haecceity began. An assemblage of the quarry, the vortex of air, the void left by quarrying, the ambience or spirit of the place, the damp leaves, Phil and Helen; and it ended as the leaves tumbled to the ground at Helen's feet. We could rationalize it as the result of a breeze we failed to notice, or we could romanticize it as the manifestation of the ghost or goblin of the old quarry, but it makes more sense to us to interpret it as a fusion of all such things – for we are repeatedly participating in such fusions.

The haecceity is an important concept within the making of diagrammatic performance. Deleuze and Guattari say that each assembled element of a haecceity is as important as another; so the context and location of an event should not be thought of as "decor," or as a "backdrop that situates subjects"; chiming with the counter-principle of "specificity" in performing with place (Smith, 2019, pp. 86–87) as a means to resist separateness and fragmentation.

Similarly, Rachel Hann argues that the attentive situation of a stage seeks to confirm a world as only being observable "in pieces, in slices" (2024, p. 83), and she warns how stages "enact and score a spatial ideology that defines and segments place through demarcation" (p. 84). The geographer Doreen Massey equally cautioned her readers that a slice – whether through time or space – is as likely to be "full of holes, of disconnections, of tentative half-formed first encounters" as exemplarily even or universally representative, warning against thinking of any space (whether the slice of a theatre stage or the rolling hills) as a "completed simultaneity in which all interconnections have been established" (2005, p. 107). With all this in mind, we have been seeking ways to create a diagrammatical performance that deliberately engages more-than-human agencies (including spaces) and does not reduce places to "scenery." Rather, it should be a collaborative endeavour with somewhere that is "being made, and always.... unfinished" (p. 107); a project of alliance and facilitation in which "the site can be its own generator of performance structure through journey, layer and unfolding" (Smith, 2017, p. 163).

Embodiment

As we have asserted above, crucial to our "being there" is walking. While we have used trains and a car to access places, it is when we get out and walk that we can "be there" in, and as, our bodies. This is not a nostalgic reaching back for some primeval presence. It is about speed and attention; we have often noticed things from a car or train, and at times we have even deployed the "car *dérive*," which involved "following the sudden impulse to veer off course, accessing the quietest roads that allowed for the slowest speeds, and stopping on a whim" (Billinghurst, 2019, p. 263). Indeed, contrary to our expectations we have found that under certain circumstances the "car has potential as a tool to map geographic zones of interest, and to pick up on, and wind into, states of hypersensitivity one might assume are accessible only through walking" (p. 263). However, it is hard to connect and reconnect, or even signal attention, when travelling at high velocity. It is also partly about touching; without a vehicle, we can touch and be touched. There are advocates for experiencing place by all forms of travel (Wilkie, 2014), and there are cautionary arguments against romanticized ideas of walking as universally accessible and equally comfortable (Rose, 2020); so our attraction to walking is due to its temporary – for us, for the time being, so far – effectiveness for a practice of "thisness," of haecceity.

While both of us have previously used walking as an autobiographical tool, enabling "the subject to relive his/her formative experiences of the world" (Lavery, 2009, p. 51), we propose here a different practice of walking as more than a method "to tease out the strange song of self" (p. 49). While acknowledging the potential for walking to "out" inner subjectivities and summon forgotten memories, we want to invite this traffic to flow both ways. So, where there is an "outing" there will also be a space – a vacuum even – left for an "inning"; for the world to rush in.

We thus offer hypersensitized walking as a practical way to think in relationalities. We can take advice from Jane Bennett to "twist and tweak usual habits," extending our bodily awareness "as a deliberate, carefully contrived experiment" (2001, pp. 24–25). Bennett's experiment draws on a "programme" advocated by Deleuze and Guattari to build the "Body Without Organs" (1988), "opening the body to connections that presuppose an entire assemblage, circuits, conjunctions, levels and thresholds, passages and territories and deterritorialization measured with the craft of a surveyor" (Deleuze and Guattari, 1988, p. 177). We push for two-way lateral movements; a finding and following of lines of desire that mean cultivating an "idea of self, not as a closed entity, but as a more open ended 'place of meetings'" (Tufnell and Crickmay, 2004, p. 290). The impulse is transversal, not transcendent; earthly rather than heavenly. (At the end of this section, we further explore an example of this kind of rhizomatic, diagrammatic transmission on Lee Moor in November 2019.)

Headlessness

Simon O'Sullivan talks of "a mode where one is more in the world" or perhaps "less extracted from it" (2024a, p. 35), that brings on "this other kind of perspective that de-centres the self" (p. 37). He suggests a sacrifice, brought about by "laying one's head on the block," and how this is required to "explore a mode of being in the world without a self" (p. 33). The payoff for this new perspective, this state of "headlessness" (which brings with it the understanding of the self as a fiction), is that it offers an opening to another kind of "complex bodily intelligence" with potential for "other heads (or non-heads perhaps)" to replace the original (p. 39). Similar ideas have been explored by Susan Blackmore (2003) and D. E. Harding (2013), while Donna Haraway also points to such a sacrifice, "the blood dripping from Medusa's head," as a useful way to situate our human-centric selves differently within the world, suggesting "we turn away from heady facial representations, no matter how snaky" (2016, p. 54). While Hélène Cixous, in her seminal essay *The Laugh of Medusa*, urges: "Your body must be heard. Only then will the immense resources of the unconscious spring forth" (1976, p. 880).

Haraway suggests that "we become-with each other or not at all" through a strategy of speculative fabulation. She advocates a "material semiotics" (we discuss such a lexicon below in Games, rituals, lexicons) that is "always situated, some place and not no place, always worldly" (2016, p.54). Similarly, Simon O'Sullivan suggests that fictioning through "writing and art practice" may be pursuits "where some of this kind of experimentation can take place," but also advises caution, for the aim is not complete self-annihilation:

> Seeing the self as a fiction is then perhaps less about identifying something that needs completely dispelling [...] but rather recognising that the device – a complex fiction that performs many important functions – is something that can be tinkered with at least to some extent.
>
> (O'Sullivan, 2024a, p. 39)

While keen to advocate for "a more decentred perception of human agency" (Kroijer and Rubow, 2022, pp. 375–384), we also suggest "necessary caution" (Deleuze and Guattari, 1988, p. 177) may be required. According to Jane Bennett, the "making strange, or 'deterritorialization' of bodily experience" requires "precision in planning, measurement, and implementation" while it is vital to "resist the lure of total freedom and wild disorganization" (2001, p. 25).

Caution required

At the beginning of the Covid-19 lockdown, Helen spent long periods alone, exploring the complex of muddy, estuary rivers on the border of Devon and Cornwall. One evening, after walking the banks of a visibly polluted urban creek situated next to the local rubbish incinerator, she was struck with a violent bout of vomiting. This experience, accompanied by some strange kind of fever dream, felt intimately connected to the polluted river: an embodied, liquid gushing that left her empty and drained. As lockdown continued, Helen continued to walk the rivers. When not walking, she was alone in her studio, painting and writing about the rivers. She talked of feeling as though she had "drained into the landscape," that she was "empty," she felt she had no self, or that there never had been a self; "just rivers and mud."

If we consider this scenario in terms of our discussion above, we could say that on this occasion, "precision in planning, measurement, and implementation" (Bennett, 2001, p. 25), was perhaps not quite precise enough.

A practice of walking, writing, and making that usually allowed for intense entangling with the landscape with little negative impact, became intensified by the isolation of lockdown leaving Helen de-centred (decapitated), but it also super-connected her viscerally and corporeally to the places she had been forensically exploring. The pay-off for the violently physical draining and sense of mental and emotional emptying came with a new (if temporary) appreciation of herself as "rivers and mud," an experience that was nevertheless useful in terms of making paintings about rivers.

As human becomings, we acknowledge that speaking for the non-human – be it fox, trout, underground lead lode, or polluted river – has potential to be at best misguided, and at worst to shore up the human-centric discourses we are seeking to dismantle. However, our making emerges from our taking time (a great deal of time) to place our bodies within the places we write about. In "Navigating Voices" (2005), Wallace Heim writes,

> Those experiences – the immediate meetings between the human and the other-than-human beings, entities, landscapes – can be likened to conversation. The relations experienced and expressed are like dialogues, reciprocal exchanges within the variations of communicative abilities.
>
> (pp. 199–200)

In this way, a conversation "taken as the whole of embodied comportment, responsiveness and communicability" rather than "confined to rational, verbal articulation" allows something mutual to emerge. This notion of emergent, intracorporeal conversation between human and more-than-human can "be extended also to the experience of meeting a work of art" (Heim, 2005, p. 200). We explore this kind of transfer below, using a chain of Crab & Bee artworks and performances that began with a simple, improvised action at Lee Moor.

The wandering voice

Playful, creative writing done on site, or immediately after a walk while the sensory experience is still fresh, can help us to reflect on the tacit encounter, making explicit the embodied implicit; "walking and writing are intimately connected: one leads to the other" (Lavery: p. 42). There is potential for a greater gestalt to be illuminated through a speculative writing process by focusing retrospectively on each agency within a place at a given time. We may try a group approach, writing together as a single voice (a haecceity). The use of pseudonyms can also help us to shake off the tyranny of a fixed or centered voice, allowing something more mercurial and slippery to make an appearance. Writing individually, we may allow a narrative to slide between first, second, and third person, between a singular or plural, or shift between tenses.

Occasionally, a "wandering voice" appears to take on its own autonomy. One begins to write from the perspective of the embodied human, but the voice takes flight, travels suddenly and unexpectedly like a bird into the world. Now (for example), it is situated in the quarry, now it speaks as the hands that clap together, now the puff of air that stirs the damp leaves, now the rising, swirling column, and finally, the whole ensemble with Crab & Bee in the quarry amongst it all.

Games, rituals, lexicons

It is 2019, we are walking on Plymouth Hoe with ornithologist Tony Whitehead and storyteller Lisa Schneidau. It is a stormy day, so we go down to the water to take a closer look at the waves. Phil produces a ball of red wool from his pocket, we cut off a thread three or four metres long. With very little discussion we throw one end of the thread into the water while holding on tightly to the other end. The holder of the thread can sense tugs, swirls, currents, and undertows beneath the surface of the water that are surprisingly different from the movement visible from above the waves. It is an exhilarating game, everyone takes a turn. We name this game "Sea Kites." Eventually, we take our soaking, briny string out of the water and hurl it at the sea wall, where it sticks like spaghetti; we tell each other stories based on the "line drawing" made by the string on the wall. This game, we call "String Scrying."

Simple actions and improvised games on site have the potential to open ourselves and others to a place, its forces, and its flows. We live in an industrial, military city where, in the name of conviviality and inclusivity, it can be better to invite participants to play a game rather than "participate in a ritual"; although the mode and outcome of both such participatory actions may be the same. In line with Haraway's endorsement of a "material semiotics," we have tied scraps of woolen thread to trees and fences, used China clay powder with a found battered dustbin lid to stencil a "sunwheel" on the grass of a community orchard, read a poem about plastic to the estuary outflow of a sewage works, created a shrine in the tiny cave of an urban beach, and performed wolf weddings in a hilltop copse by a tungsten mine on Southern Dartmoor.

Such improvised actions may subsequently be re-enacted with others; inviting guests to walk with us and performing for/with them in the place where the initial action emerged. Or the initial improvisation may be worked into a score, which we then pass on for others to try. Or extracted from the site altogether and worked into a longer and more developed performance in a "non-site." While such actions are often initially spontaneous, unplanned, and made in response to a place, we advocate assembling a portable kit to carry for such occasions. What you put into the carrier bag will vary from person to person (and from site to site), for, as Chris Crickmay and Miranda Tuffnell suggest, "a web of connection grows out from what we know of the physical qualities and behaviour of any material – a network of stories, instances and applications that enlarge our sense of what that material is or means to us" (2004, p. 166).

During our walking, we have noted patterns and codes in the English landscape that permeate not just folk traditions but also more formalized sports. We have observed players in white on green grounds engaged in games of cricket, bowls, tennis, and croquet, as well as maypole and morris dancing. The language used across such activities – between arts and sports – is often interchangeable. They share the same historical origins in traditions of "pastimes"; festive pursuits enjoyed during high days and holidays. Teams in cricket and morris dancing are equally referred to as "sides." The green space of a cricket ground or a bowling green is scaled down but still evident in the green baize of a billiards or card table, while patterns within both games are evident in the relations between the pegs in the board game Nine Man's Morris.

While much of this coding tacitly props up problematic structures and jingoistic ideas of "Englishness," by making ourselves aware of these implicit diagrams we can attempt to unpick and even "redraw" the rules and codes to undermine such structures. For example, the black tatters of Dark Morris and Border Morris shift the anachronistic tone of a sentimentalized Olde England to something far more macabre and subversive. The Cambridge-based Molly side Gog Magog wear rainbow clothes and brightly painted faces, side-stepping complicity with racist parody (which is possibly part of Molly's history of face-blacking), overtly rejecting "divisive white supremacist and white nationalistic ideologies, and any attempts to co-opt our tradition for racist ends," and taking an inclusive stance (Gog Magog Molly website, n.d.). Gog Magog's affiliation to the patterns and flows of the landscape is redrawn in their new/traditional dance *Reality Checkpoint*. The title and content of the dance comes from a famous lamppost of the same name located in the middle of a large green in Cambridge city centre, which is used for sports and festivities throughout the year. Gog Magog's *Reality Checkpoint* diagrams the traffic of cyclists and pedestrians over the paths that criss-cross the green, repeatedly meeting at the centre; a pattern immediately recognizable by those with local knowledge.

The absorbing and rejecting of different cultural and social nuances has always been part of folk "traditions," which may be why it has been so difficult for scholars to unpack what elements of which folk traditions derive from what past. Part of a teasing out of patterns and codes, and working out how to use them, is to identify the materials you can work with. Then create from them a lexicon of interweaving narratives, leitmotifs, images, and/or colour coding. Helen cites an example from her *English Diagrams* exhibition in 2018:

> I came to understand that a personal compulsion to pick up white chalk and carry it back to my studio was connected to memories of

my childhood in the chalky landscape in East Anglia. Equally, that the red earth of Devon has associations with my later life in the West of England.

(Billinghurst, 2019, p. 273)

Materials may thus be selected as "points of departure for interweaving stories that can be set to work during studio production." A lexicon draws together "colour, materials and even spatial direction and temporal indication" to support a practice. In this way, it informs the "essential elements of an artist's creative language, and also set[s] parameters for the work made in the studio; like the rules of a game that only unfold as it is played" (Billinghurst, 2019, p. 273). Once established, this lexicon serves as a basis for future work, and you can continue to develop it with every place you visit, and each performance you make.

Lee Moor, November 2019

Crab & Bee make a visit to Lee Moor; the site of a vast complex of working China clay pits (and all the ecological devastation that comes with this) on the edge of Southern Dartmoor. Two black crows fly out of a pine forest and over our heads as we climb the green hill towards the edge of the pit. We gaze down into the immense creamy white cavity at the top of the hill; a science-fictioned, alien landscape. A Land Rover cruises slowly by; two security guards sitting inside eye us suspiciously. We ignore them and continue to walk slowly along a bank of green grass, a barbed wire fence marking the pit perimeter on our right. The men in high viz get bored and drive away.

A jumble of white sheep bones is scattered on the mossy bank at our feet. There is not a scrap of sinew on them: the crows have clearly feasted and the bones have been there for some time; some of them have grown a little green in places from the rain and lichen. We pick up some vertebra to examine: each bone is penetrated by a heart-shaped hole. Phil pulls two balls of wool out of his pocket. Without discussion, we thread two three-metre-long strands of wool – one red, one white – through the heart-shaped holes of seven of the bleached backbones. We each hold the ends of two threads, then move apart; the bones tinkle together as they dangle between us. We swing the assemblage gently, side to side. Once, twice, three times, and then over like a skipping rope, and the red and white threads twist together like a candy cane.

When the action is done, we tie the twisted threads to the perimeter fence. It feels something akin to a memorial, a message to horse riders on the adjoining bridlepath; an acknowledgement of the exterminated ecology and devastated landscape we are standing in (see *figure 5.1*)*. But perhaps also some kind of unspoken spell; a jinx on the corporations enacting the large-scale damage we have witnessed. We put seven more backbones into an empty lunch box and take them away for further use.*

Figure 5.1 Helen Billinghurst of Crab & Bee ties threaded sheep bones to a barbed wire fence on Lee Moor, 2019. Photo: Phil Smith

The action described above was improvised entirely without discussion or pre-planning. It was made in the place with materials found on site, and other materials (red and white wool) we had brought with us in our carrier bag. On returning home, Helen wrote a poem about the event:

Lee Moor

1 Two crows.
A sheep's bones picked clean
on the grass
by the claypit.
A painting, or perhaps a game:
red
removed from
white
by black
on green.

2 I'll sing you a song of love and death
desire and stillness,
heart and skull.
A sing-along song of red and white
West and East
blood and bone.
I'll sing you the swing-along song of us
sheep and crow
spine and moss.

3 Two threads
strung
with
back bones
whirled:
a skipping game,
a horizontal maypole,
a blessing,
a charm,
a warning,
a curse.

Helen emailed this poem to Phil with the suggestion that it could function as a score or set of instructions for a game: "red inside black" she wrote, cryptically. Shortly after, following her own advice, she used the poem as a "score" to construct a painting of a crow. First creating a green ground, she laid down white paint, then red paint, and finally a layer of black paint to create *Crow* (2019).

Bones swinging

The action of "swinging the bones" combined with reading the Lee Moor poem quickly became a strange, slightly macabre part of the Crab & Bee repertoire. First performed at poetry readings, then as part of our *Plymouth Labyrinth* (2019) project, we swung the bones with and for audiences (human and non-human) inside performance spaces, and outside in the field, from one side of the country to the other. The painting *Crow* (2019) – made initially to hang on the wall in our Plymouth Labyrinth exhibition – was conscripted and redeployed in later diagrammatic performances. Paraded across the performance space by Phil in *Webs, Nets and Carrier Bag* (2021), *A Séance for Stonehouse* (2023), and *Patterns, Diagrams and Place-Stories* (2024); the *Crow* painting traces a line of flight or thread of story across a "landscape" marked by a web or pattern of other colour-coded journey-lines. In this way, a simple, improvised action, made as a deliberate act of human/

more-than-human corporeal entanglement, spun out diagrammatically; moving across times, spaces and bodies, traversing a range of media and shifts of scale.

Stylistically and formalistically, none of this is anything new in performance practice. It is neither ahead of the curve nor an attempt to get ahead of it. It certainly fails any fashionable test. Just as many of our allies are pieces of the system's detritus, trash, so most of what we are suggesting here, and everything we have used, are discarded materials from previous avant-gardes. In fact, we fit rather comfortably into Richard Schechner's description of a theatrical "tradition-seeking avant-garde" with a "stripped down stage, the ritualized nature of the encounter between performers and spectators, the startling confrontation of extremely personal expression and totally composed face and body 'masks'" (Schechner, 1993, pp. 11–12). However, our practice is not a discipline of physical training, but a routine of attention to an abject terrain, which shifts the focus of our encounter, displacing it from a space between performers and spectators, to one of three dimensions; a complexity.

Fictioning

Fabulous images versus memes

Mark Fisher's rallying call, for an alternative to the "widespread sense that not only is capitalism the only viable political and economic system, but also that it is now impossible even to imagine a coherent alternative to it" (Fisher, 2009, p. 6) was a catalyst that set theorists and artists scurrying towards fictioning, fabulation and speculation in attempts to "tear a hole in the grey curtain of reaction" that had "marked the horizons of possibility under capitalist realism" (p. 85). Post-pandemic, the neo-liberal impasse appears to have shifted; maybe not in the way hoped for by progressives, but further towards global reactionary plutocracy. With "[t]erms such as 'alternative facts,' 'post-truth,' and 'fake news'" as "overnight clichés frequently used to discredit oppositional voices," articulating some version of a post-rationalist relativism, perhaps it is reasonable for the network theorist David Garcia to ask "to what degree were we ourselves complicit in the emergence of this new toxicity in public discourse" (Garcia, 2017, p.78); a discourse fuelled by the "dark humour and love of transgression for its own sake" of "meme-making trolls" (Nagle, 2017, p. 2). Indeed, meme culture on both left and right share common roots in the sigil-writing of late twentieth-century chaos magicians, and to the (initially left-leaning) anti-austerity cyber-activism of Anonymous early in the twenty-first century.

The "complex online ecology of 4-Chan message boards" (Garcia, 2017, 96) and the rhizomatic spread of conspiracy theories may seem – in structure and shape – similar to the fictioned webs we weave across the landscape. Like the meme-makers, we also deploy, in puckish mode, "the scrambling of already-existing code or the importing of more alien code from elsewhere." For example, we may assign new meanings to a corporate logo encountered while walking, which then becomes part of our own lexicon, or perhaps we recruit a local saint to join our canon of personalities of place. By following these methods, "an art practice [....] can then take off from this hybridity and begin to work on its own terms, producing its own (autonomous) coding" (O'Sullivan, 2016a, p. 3).

Despite these methodological similarities there are key differences between the mode of fictioning within diagrammatic practice that we advocate, and the modes of the online meme-makers. Firstly, our tangled webs are woven across the *physical* landscape in response to sensual information; the images that speak back to us are usually from "the outside" rather than received through the human algorithmic echo-chamber of a screen. Secondly, our transmissions are inter-corporeal: we use affective, haptic processes and materials to transmit ideas and images amongst human and more-than-human worldings. Thirdly, we do not hold "the accelerationists' passion" for the human-centric and "hyper-compressed discourse of meme and tweet culture" (Garcia, 2017, p. 97), preferring to play a longer game, to work at different speeds and slownesses. We return to sites, repeat patterns, and re-imagine images. We develop discourses in a way that celebrates loops, spirals, and long-circuitry. We generate new/old images and mutant motifs that step forward from our repetitions.

Like the writer William Burroughs, we "look outside, to achieve as far as possible a complete awareness of surroundings." Burroughs, famously, sought ways to cultivate a kind of non-verbal, pre-linguistic "thinking in images"; a mode to correlate the "outside" with the "inside," to bring attention to a "psychosensory process that is going on all the time anyway" in the form of "extraordinary juxtapositions" (Callas, n.d., unpaginated)

> For example, a friend of mine has a loft apartment in New York. He said, "Every time we go out of the house and come back, if we leave the bathroom door open, there's a rat in the house." I look out the window, there's Able Pest Control.
>
> (Burroughs and Gysin, 1978, unpaginated)

Resonating with Deleuze's conceit of "The Fold," and Jung's discussion of synchronicity, Burrough's world of images syncs with the sensual world: "For Godsake, keep your eyes open. Notice what's going on around you" (Burroughs and Gysin, 1978: unpaginated). In developing the cut-up technique, Burroughs sought not only to dissect, fold, and re-arrange words on a page, but to generate the new by "breaking down" what he referred to as "the control system" (Burroughs and Odier, 2008, p. 33).

"Fabulous images" and motifs that stand forward to an author or maker may operate as "'key points' in the production of subjectivity" (O'Sullivan,

2024a, p. 16), but it is worth stating that the images that step forward for us are frequently in physical form. For instance, a rusty key found on the path next to the Wolf Minerals tungsten mine becomes St Hubert's key, used historically (and now within Crab & Bee's performance *Leadworking* [2023]) to cure (care for) or conjure werewolves. Another example is the space rocket-shaped broken trowel found alongside a flying saucer-shaped piece of discarded plastic in a cobbled back lane, early in our collaboration. We had no placing for these celestially twinned objects within our lexicon when we found them, but so striking were they that Phil continued to collect such "space junk" during our forays. Five years later, this box of collected "spacejunk" became the ritual reliquary for our *Scoriton Mysteries* performance (2025).

Uncovering

Another example of the emergence of an important motif in physical form, is artist Kemang Wa Lehulere's chance discovery of human bones, when digging through earth with an afro comb for a performance in Gugulethu, Cape Town, South Africa. This chance unearthing continues to resonate through a body of work made twelve years later. The initial performance, *Ukugaqula iBatyi* (2008), was made "within the context of an exhibition called *Scratching the Surface* as part of artist collective Gugulective." Much of Wa Lehulere's subsequent lone and collective work still interrogates this unexpected discovery. His work exposes historical cultural erasure by "combing" through the archive, while using creative writing to interrogate "the importance of oral history" and "unwritten narratives" (p. 60). Wa Lehulere excavates post-apartheid South African pasts and futures by continuously looping back to the spontaneous moment of his combing through the earth of Gugulethu (Dyangani and Lehulere, 2017, pp. 43–68).

So, when referring to diagrammatic performance-making we are always discussing embodied, interdisciplinary processes; even in "thinking as well as making practices" and in "pedagogical practices and cosmological performances" (Haraway, 2016, p. 14). We have come – albeit in the non-neurotypical mode of the synaesthete – to think increasingly of story as material that is more of substance than abstraction; as the connective tissue of pasts, futures, and places.

The practice of fictioning – or supposing – has served us in "cultivating new ways of imagining the landscape" (Billinghurst, 2019, p. 236). This re-imagining can involve a performed re-arranging and re-ordering of the image/objects of a lexicon; a re-diagramming of worlds comparable to Burroughs' cut-up; "systematic and repetitive, simultaneously disconnecting and reconnecting—it disconnects the concept of reality that has been imposed on us and then plugs normally dissociated zones into the same sector" (Burroughs and Gysin, unpaginated). It also operates as a physical version of O'Sullivan's discussion of the "recontextualization and manipulation of concepts" as if in a game "on a table-top" allowing "hitherto 'illegal' connections and syntheses to be made" (O'Sullivan, 2016b, p. 21).

In a performed curating (from the Latin *curare,* meaning "to care" or "to cure") of image/objects, and the re-drawing of their relationalities, we can also draw on Ursula K. Le Guin's "narrative conceived as carrier bag/belly/box/house/medicine bundle." Indeed, the materials and image/objects of Crab & Bee performances are often produced from actual supermarket shopping bags containing the "necessary elements of a whole," the purpose of which is "neither resolution nor stasis but continuing process." Doing our utmost to undo heroics, we eschew the "killer story" (Le Guin, 2019, p. 168) with its "proper shape" starting "here and going straight there" (p. 169). We prefer winding routes and tangling yarns. We resist tales of the "triumphant (Man conquers earth, space, aliens, death, the future, etc.)" as well as the "tragic (apocalypse, holocaust, then or now)" (p. 170).

Everyday entities

Staying with themes of winding routes and re-imagined landscape, we can shift attention to the very real agencies of "fictioned" entities on global civic engineering, infrastructure, and road planning. Examples of the impact of such "non-entities" include the mass-media controversy and re-routing of Newmarket-on-Fergus bypass around the "fairy tree of Latoon" in County Clare, Ireland (Mulvaney, 2019: unpaginated), there are the re-arranged road plans in New Zealand to protect a *taniwha* (a protective spirit of the Ngati Naho people), and finally the protest over four proposed Icelandic roads threatening the habitat of the *huldufolk* (hidden people) (Reeves-Evison and Shaw, 2017, p. 16).

Theo Reeves-Evison and Jon K. Shaw suggest that such surprisingly common scenarios are rarely as simple as "authorities giving ground to a product of make-believe," but that it is also not right to "attribute a devout faith in *huldufolk* to Icelanders" (p. 16). The reality, far more nuanced, is that "the majority of local inhabitants do not so much believe in *huldufolk*" as like to "entertain a belief in them" (p. 16). Reeves-Evison and Shaw suggest there is a "powerful desire" amongst Icelandic people to "preserve such geological platforms for their imagination." In other words, the land is "speaking back" to the people and the license to entertain belief in the *huldufolk* is not "opposed to the physical materiality of Icelandic rocks but entwined with them" (Reeves-Evison and Shaw, 2017, p. 18). Discussion around the "existence" of fairies is a distraction from their potential to advocate for the more-than-human, and to address the "need for 'new' kinds of figures and images that are appropriate and adequate to 'living in the ruins'" (O'Sullivan, 2024a, p. 142).

Such examples demonstrate the "real effects" of entities "unknowable to materialist science" (Reeves-Evison and Shaw, p. 18). Reeves-Evison and Shaw suggest that "there is an important role for both material location and for continuity and repetition in maintaining these fictions as powerful operative forces in the world" (pp. 19–21). Emigrants removed from Iceland and living abroad are "less inclined to believe in much of their country's native folklore," and the fifty-year New Zealand Tohunga Suppression Act (1907–1962) was "enough to break uncountable threads in the passing down of *tohunga* wisdom." On the other hand, a belief in local spirits "equally takes on more and more strength through iteration" (Reeves-Evison and Shaw, p. 21). Such fictioned iterations and repetitions would thus appear to sharpen human "supposing skills"; for entertaining the possibility of sentience and personhood in non-human others, and for questioning powerful and pervasive extractive discourses. A step, perhaps, towards the granting of legal personhood (and rights) to rivers, forests, and oceans. As such, fictioned Irish *sidhe*, the *taniwha*, the *huldufolk*, and the *genii loci* of all other places still hold powerful agency as interlocutors between human and more-than-human for future times, as they have in times past.

Changing the ecological cast of characters (not ghosts but goblins)

Agencies

The task we have set ourselves here – for which we are drawing on diagrams and site-specificity for assistance – is to advocate for a performance that invites more-than-human agencies (as "defined," in part, in the section above) to become present as co-performers. To gather the means for a performance that re-establishes embodied presence over appearance and representation. The challenge for human performance makers, given the deathly structural calibrations of relations inscribed within representation itself, is where to make (or find) a space to create work that values material things in their own right. Not as "material" for a representation. How to make performances that do not collapse matter into images but shift the dominant relation in a performance from one that privileges images to a flattened diagrammatical ground shared by its human and more-than-human agents. At the same time, there is a need for extreme wariness around affirmative ideas about nature and matter – of, say, immersion, eco-sexualism, nature spirituality, the healing of the Earth, and so on – not because such content is problematic (though it may sometimes be), but because any representation that is not undoing itself is very likely to be an invitation to become part of the representations.

So, how do we resist the tendency to dematerialize nature while working in an art of representations?

The first part of our answer is to always shape our earliest preparations in embodied presence. While we have no claim on any space immune to the operations of spectacular relations, we have taken the tactic of the *dérive* from the situationists (destination-less wanders away from cultural centres) in order to get bodily to post-industrial ruins (Edensor, 2005), edgelands

(Shoard, 2002), and voids (Careri, 2002) where the mechanics of the spectacle are weakest or have tended to short-circuit themselves under the stress of economic crisis. In the UK, these are unlikely to be found "in nature" (given the preponderance of a spectacle of managed forestry, industrialized farming, and national parks) but are more likely found in ambiguous interspaces where managements lose track, functional distinctions break down, and boundaries are blurred.

During preparatory (research) visits to these spaces, our intention is to pay close attention to the shaping and textures of materials, to the rocks, soils, and the more-than-human animals, plants, and fungi there. At the same time, we are focused on ambience as much as we are on data. We are particularly attentive to the place-stories: stories and legends that are self-evident, signed, or emerge from a place. These we take to be agents in (as well as accounts of) a place's own making of itself. We treat the objects or things of the place with the same respect. We regard them as being as material and agentive as the place's stories; sometimes we may extract them (on a very small scale: a bottle of water, a handful of bones) to a non-site (Smith, 2025, pp. 62–67) of performance, before (where possible) returning them to where we found them.

Affordant entities

One of the key tactics of the situationist *dérive* is to navigate by sensing out affordant ambiences: those atmospheres that invite a walker to a place. While this might seem too much in tune with a dematerialization of relations in a society of the spectacle, at the same time (for everything in our performance-making is made in tactical compromises with, and within, a totalizing culture) such a sensitivity to ambience gives us access to those stylistics of resistance in hauntological art that retain a longing to push into the future: a "materialist sorcery" in which "the seam" of its production is "turned inside out" for us to experience (Fisher, 2014, p. 90). Thus, in the places to which we are drawn, we can work through their atmospheres and other immaterial presences to get at the matter of those presences (the signs, the stories told in the pub, the church magazine, a discarded trade union banner in a hedge, a dried up river bed, shrapnel rusting in pavement slabs), not because one is more real than the other, but because, in Mark Fisher's words, "it is only when the possibility of supernatural spooks has been laid to rest that we can confront the Real ghosts… the ghosts of the Real" (p. 107).

Fisher's mention of "the Real" here, is not a general address to "reality," but refers to psychologist Jacques Lacan's term, glossed by Benjamin Noys as "an experience of shattering enjoyment that lies outside the field of representation" (2016, unpaginated). While it is potentially daunting in its intensity, "the Real," like a "haecceity" (see above), holds out a promise of something that escapes the tyranny of images. So, while we may entertain the various fanciful elements of a place-story, it is to "lay them to rest" to get through to the sensorial overload that they are holding back.

Jacques Derrida derived his term "hauntology" from the opening line of Karl Marx and Friedrich Engels's *Communist Manifesto* of 1848. This is usually translated from the German into English as "a spectre is haunting Europe – the spectre of communism." Derrida proposed that it was now in the nature of our being human, alive in the aftermath of the unrecoverable ruin of communism's alternative to capitalism, to haunt ourselves. There has been some usefulness for us – both in making performances and in preparatory visits – in drawing on artworks often described as "hauntological," given their knowing or accidental redeployment or *détournement* of the ruined representations of failed utopias as critical devices. A contemporary example of such a *détournement* is Libby Bove's *Museum of Roadside Magic*; a travelling collection of motoring symbols and signs – MOT motifs, plastic bollards, traffic signs – that Bove has repurposed as the sigils, costume paraphernalia, and oily wands of a combustion engine paganism. With her van and trailer-museum, motor-based bespoke tarot pack and t-shirts, Bove straddles the festival and gallery scenes. She blends her found roadside objects – bits of engine, car manuals, fobs and ignition keys – with folk song and morris dance, to create an imagined folklore around annual road safety tests (see *figure 5.2*). Rather than speed – the elixir for accelerationist Futurists – it is a lost regulation that Bove celebrates. While immersed in the diesel, petroleum, and oiliness of repair shops and garages, Bove fills her exhibits with the aroma of roadside herbs. There is nothing antiquarian, quaint, or "precious" about her invented nature religion made from the waste products of motoring.

Libby Bove's vision is of an imagined society that has never persecuted nature religion; in which practical everyday things are permeated with herbal lore, in which the shamanic and the mechanical are identical. Her living montages of *mari lwyd* and *penglaz* (guising horses) in hi-vis jackets, of witches in knitted conical bollard-hats, and Masonic-style aprons decorated with spanner symbols, evoke a never-existing cultural orthodoxy tantalizingly dependent on a dominant petrol culture. Like many hauntological projects – from movies, to fictions, to newly modelled rites – *The Museum of Roadside Magic* affects a deep melancholy in response to a cancelled future and at the same time expresses a critique of that cancellation. Like the best hauntology, Bove's work also offers an intuitive vision of an alternative future; not of an overarching model, but of useful materials for a future salvagepunk (see Salvagepunk section below).

Goblins

Noting how "whilst not all of us enact.... a literal figuration of living in ruins, we mostly do have to work within our disorientation and distress to negotiate life in human-damaged environments" (Tsing, 2015, p. 131), we

Figure 5.2 "Cone Dancer, Radstock" from the series "So Turns the Wheel MOTO" by Libby Bove, (2022–ongoing). Photo: Libby Bove.

have drawn on the materials of disorientation and imaginary ruined worlds, such as from episodes of *Doctor Who* and the 1970s BBC children's TV series *The Changes*, deploying them as a lens to interrogate abject places, around disused mines and quarries, for example. In this way, we address such hauntological sites as levers pressing on present matters to break through to "the Real" and the haecceities that the places conceal.

We take our cue here from Helen Macfarlane's 1850 translation of Marx and Engels's "*Gespenst*" (from the opening line of their *Communist Manifesto*) not as "spectre," but as a "frightful hobgoblin," as an invitation to the discomforting enjoyments that defy representation in these places. The hobgoblin, with its connotations of labour, grime and materials, rather than the lifeless spectre, is a better model for the casts of characters we often encounter in the story-places of our preparatory walks. On these walks – every bit as performative as any formal performance, with ritual-like actions, embodied immersions, libations and washings – around post-industrial mining sites and quarries, rather than the ghosts of the past, we have been more aware of a goblin-like presence, not only in the stories there, but in and as the materials of these places. Associated with mining or metalwork (blacksmithing, for example) – stunted, forbidding and alienated figures shaped by arduous labour, driven by necessity far more than idealism – these goblins or Cornish "knockers" are identified not only with the work done and the worker labouring but with the materials laboured upon.

Here we can draw on a key principle from David Abram's *The Spell Of The Sensuous* based on the author's witnessing the food left out by a Balinese villager for household spirits being eaten by ants: "a strange thought dawned on me: what if the ants were the very "household spirits" to whom the offerings were being made" (1997, p. 12). So, in both our performances and in our encounters with a place-story of a goblin, mermaid or witch's curse, the same "strange thought" strikes us: what if these are not metaphors for – or anthropomorphizations of – a mine, a stream or a poisoned underground river, but they *are* the mine, the river, and the cavern, and our entangling with them. Working with such "Real" characters – "by propelling [our] awareness laterally, outward into the depths of a landscape at once both sensuous and psychological, the living dream that we share with the soaring hawk, the spider and stone" (Abram, p.8) – we are not making performances *about* places, but synergetic diagrammatical place-performances and performance-places. Hauntology is thus a staging post from which to reach out to the things of the mine, river, and winds.

Hyperobjects and vibrant matter

Connected

Since Rachel Carson drew dots together to connect public awareness to a previously little noted systemic poisoning of water, air, soil and food webs by industrial farming methods using combinations of chemical pesticides in *The Silent Spring* (1962), many others have followed in her wake, not only adding to the empirical evidence, but broadening the conceptualizations and re-focusing the perceptions of environmental depredation. Posthumanist Rosi Bradotti, for example, argues that "[l]iving matter – including the flesh – is intelligent and self-organizing, but it is so precisely because it is not disconnected from the rest of organic life" (2013, p. 60), while posthuman feminist Astrida Neimanis, has traced out toxic elemental flows on a huge scale through earthly and creaturely "bodies of water"; monitoring, in particular, the trickling transfer of "planetary breast milk" (2017, p. 36). Donna Haraway, in turn, argues that mainstream discourses fail to address the entangled and expansive reach of these phenomena, suggesting that the "sciences of the Anthropocene" are "too much contained within restrictive systems theories.... namely the old one of bounded utilitarian individualism" (Haraway, 2016, p. 49).

Vast scales and the long durée

Philosopher Timothy Morton defines such vast systems as "hyperobjects" (Morton, 2013). Hyperobjects are "massively distributed in time and space relative to humans" making them, in turn, hard for us to perceive and/or respond to; being "viscous," "non-local," and performing at "profoundly different temporalities than the human-scale ones we are used to." Hyperobjects include giants such as Covid-19 and "Styrofoam or plastic bags, or the sum of all the whirring machinery of capitalism" (p. 1). Morton suggests our increasing ability to detect such systems, operating on epic spatiotemporal registers alongside the comparative insignificance of the singular human life-scale, has "already ushered in a new human phase of hypocrisy, weakness, and lameness" (p. 2), as evidenced by the bizarre and frightening inadequacies of current geo-politics.

Thinking and performing on and in relation to such scales also presents considerable challenges for artists wishing to engage with or respond to the hyper-object, but some do attempt this. For example, in *Ping 10* (2010–2011), movement artist Simon Whitehead and sound artist Barnaby Oliver held a year-long online creative dialogue, responding in and to their respective local rivers; the Cych in west Wales and the Maribyrnong, in Melbourne, Australia, "a few thousand miles apart." In this work, they approached the two rivers as an "unbroken body of water through which they would correspond" (Oliver and Whitehead, n.d., unpaginated).

In 2019, Icelandic-Danish artist Olafur Eliasson returned to Iceland to photograph glaciers he had first photographed twenty years before. "Suddenly, things.... are shifting speed" said Eliasson (2019), shocked by the dramatic changes he found there. His assumptions about the rate of the impact of climate change on glaciers he had previously considered to be "outside human space and human time" were radically challenged. At the end of a year spent re-photographing the vanishing glaciers, Eliasson accompanied poet Andri Snær Magnason and geologist Oddur Sigurðsson to the site of glacier-no-more Okjökul, declared to be "dead ice" by Sigurðsson in 2014, to install "the first monument to a glacier lost to climate change anywhere in the world" (McCraig, 2019, unpaginated).

The ability of the hyperobject to perform over unthinkable timescales is the subject of the documentary *Into Eternity* (2010), directed by Michael

Madsen, about Onkalo, a deep geologic repository for nuclear waste in Finland. The rock selected for this aeonic burial is 1.8 million years old; once the specially made cavity is filled to the brim with its lethal cache, the intention is to seal the site and leave it undisturbed for a hundred thousand years. "At the surface, the clock runs very fast, while in the rock it goes slowly," says interviewee and research director Peter Wikberg. *Into Eternity* grapples with how best to prepare for this, asking "how do you create warnings for the future over such a timescale?" Marking the site with warning signs or trying to seed a grim myth to ward away the curious would likely be unsustainable over such a timescale. Meanings, symbols, and languages change, and enigmatic myths could be counteractive, luring in rather than repelling would-be explorers. There are resonances here of the spreading of tales of "mummy's curses" associated with the sealed pyramids of Giza and other ancient tombs; ineffectively as it turns out. In *Into Eternity*, science editor Berit Lundqvist argues against such an approach. "We shouldn't have to guard it, it should just be able to be left," she says, "because conditions on the ground will change; conditions down in the rock are stable."

Artists Robert Williams and Bryan McGovern Wilson similarly address deep time, landscape, myth, and *genii loci*, exploring in their collaborative project *Cumbrian Alchemy* (2013) what the markers of toxic sites could be. Their work considers the future/past of the nuclear and mining industry on the north-west Energy Coast in relation to the geography, archaeology, and folklore of North Lancashire and Cumbria (UK). Developing the performing of a "[P]riest-like figure" drawn from the persona of J. Robert Oppenheimer ("father" of the atomic bomb) to ward off and warn future generations of the potential toxic hazards of the site:

> The character retains the hat and suit but is masked, and assumes the role of a Guardian, through whose situated performance future generations might be warned off the sites of radioactive waste. The focus is then to create a prospective memory.
>
> (Williams and Wilson, 2013, p. 39)

In creating their cautionary myth, Williams and Wilson drew from ideas developed by Hungarian folklorist Thomas Sebeok, who had worked previously for the US Nuclear Regulatory Commission "to consider ways to mark nuclear sites for future generations." Sebeok proposed seeding "narrative and oral traditions, as well as a sequence of signs to create an atomic priesthood" suggesting this as the approach most likely to withstand the *longue durée* required for such a project, rather than any monumental architecture (Williams and Wilson, 2015, p. 114).

Sobeok's conceit resonates with Diana Taylor's discussion of the "repertoire," suggesting that "embodied expression has participated and will probably continue to participate in the transmission of social knowledge, memory, and identity" (Taylor, 2003, p. 16). Geographers Patrick Nunn and Nicholas Reid's comprehensive survey of indigenous stories from twenty-one coastal locations in mainland Australia would seem to back up Taylor's and Sobeok's arguments about longevity of transmission, proposing that "Aboriginal stories about coastal drowning [...] in most cases are considered likely to recall the effects of postglacial sea-level rise more than 7000 years ago." In a global context "this raises the possibility that stories of comparable antiquity both in Australia and elsewhere in the world exist and may yield information that is likewise of interest and significance to a range of questions" (Nunn and Reid, 2015, unpaginated). Another such example of loric transmission enduring over generations are the "plausible descriptions of a volcanic eruption" told by an Aboriginal elder of the Gugu Badhun Kinrara, Queensland, Australia. Geochronologist Ben Cohen, who dated the volcano's most recent eruption as at least 7000 years ago, uncovered the recording, which was made in the 1970s:

> The elder described a time when a pit was made in the ground with lots of dust in the air, and that people got lost in the dust and died. He also described an occurrence when the earth was on fire along the watercourses.
>
> (Cohen, 2017, unpaginated)

Elemental

While we continue to float ideas about the relations between folklore, culture, and lively matter (see Sexy Theory section below), we also regularly shift scales and turn our focus to the agencies and specific properties of materials and elemental particles as storying stuff, performing inside and outside our bodies. It is important to acknowledge – given that we are performing with these materials – that the (human) body requires some amount of water, calcium, chromium, copper, iodine, iron, magnesium, manganese, molybdenum, potassium, sodium, and zinc to function. Yet most of these elements will become toxic if too much is ingested; it is a fine balance.

Lithium

Take lithium, for example. Third on the Periodic Table, it was present almost immediately after the "big bang" and is one of the lightest elements. It is now in high demand for the manufacturing of lithium-ion batteries, used for electric vehicles and storing renewable energy, and as such is seen as crucial to "combatting" climate change.

Lithium was also first used to successfully treat bipolar disorder in 1949 (Cade, 1949, p. 520), yet it is still not fully understood how this works. Its long-term use can cause numerous side effects on the human body, including multiple organ failure, and is particularly hard on the kidneys. When told she needed to change from this bipolar medication due to the impact long-term use was having on her kidneys, writer Jaime Lowe made a "pilgrimage" to Bolivia, where lithium is extracted from brines in salt flats. After twenty years of stability, having previously experienced the trauma of the extreme highs and lows of a bipolar condition, Lowe describes her relationship with lithium as "an almost love affair." She says of her visit:

> It feels like you're in a completely different universe, where everything is kind of this crystalline green and you can kind of feel the salt crawling up your body and sort of immersing itself in your pores.... I wanted to just roll around in it and kind of pay homage to this thing that had helped me for so long.
>
> (Lowe, 2017, unpaginated)

Author, musician, and podcaster "Blindboy Boatclub" – originally a member of the Irish band The Rubberbandits – draws lithium and a traditional story weave into contemporary culture through the lore of his live podcast performances. Knitting his own lived experience as a neurodivergent individual with Irish folklore, fiction, cultural critique, politics, and comedy, "Blindboy" raises discussion around contemporary mental health issues. His live events include interviews with invited guests such as folklorists, psychologists, and storytellers (Chambers, 2022). Amongst other legends, "Blindboy" has discussed the phenomenon of Gleann-na-ngealt ("Valley of the Mad") in County Kerry, Ireland (O'Connell, 2012). This valley has a tradition of orally transmitted place-stories about visitors to the region being cured of mental illness by drinking from its wells and eating watercress from local streams; both of which have been shown to contain lithium as a trace element. From the very ancient pre-Christian myth of Mish (a princess who took form as a blood-drinking feathered demon after seeing her father decapitated) to the medieval legend of Suibhne mac Colmain (Mad King Sweeney who was prone to flying into feathered frenzies), the tales continue to hold meaning and attraction for people suffering from mental illness today. Pilgrims continue to make the journey to Gleann-na-ngealt seeking a cure from drinking water from the Tobar na ngealt (Mad Well) and milk from the nearby "Mad Stone."

Lead

Another element intimately inscribing itself into human lives is lead. It is referred to by reporter Kevin Drum as "America's true criminal element" (2013, unpaginated). Multiple studies have indicated how "even moderately high levels" of lead exposure during childhood are "associated with aggressivity, and impulsivity, ADHD, and lower IQ." The studies also indicate a direct correlation between quantities of lead in petrol and "the rise of [violent] crime in the '60s and '70s and the fall of crime in the '90s" (Drum, 2013, unpaginated).

In the past, lead poisoning was referred to as "Saturnism," because early alchemists believed that lead was associated with the planet Saturn. As the heaviest of the seven "planetary metals," alchemists thought of lead as a base matter to be transmuted into gold. Because of this, and because of its malleable qualities, artists have continued to be drawn towards this enigmatic substance. Over the course of a year, Richard Serra worked through his celebrated "Verb List" (1967) as instructions to work with lead. He created over one hundred works, including a film *Hand Catching Lead* (1968) in which a hand repeatedly grasps to catch chunks of lead dropped from above the camera frame, *Thirty-Five Feet of Lead Rolled Up* (1968), and the propping of four sheets of lead against each other to create *One Ton Prop (House of Cards)* (1969). More recently, British artist Cornelia Parker melted down lead bullets, transmuted the substance into a fine wire which she then worked into paper to create her *Bullet Drawings* (2010).

Transmutation, pilgrimage, and water

Similarly intrigued with material transmutation, Joseph Beuys lined the walls of the Konrad Fischer Galerie in Dusseldorf with lead to create *Pain Space* (1993). Beuys thought diagrammatically; he thought of society in relational and sculptural terms, drew his ideas out as literal diagrams on enormous blackboards over a series of durational performance lectures, and used materials to map out flows of capital and transfers of energy and power across non/human ecologies in his sculptures and performances. His work conveyed "forces and energies of the natural world, often grasped at a prelinguistic or presymbolic level, through his personally forged language of forms and substances" (Adams, 2014, p. 2). Beuys's material lexicon included felt, honey, and fat, and he referred to nonhuman animals such as the hare and the stag as totemic power forms, rather than representations of human tropes and attributes. A crucial element of Joseph Beuys's work was pilgrimage; making meaningful journeys to, and actions in, several countries including the United States where he performed *I like America and America Likes Me* (1974), Scotland where he performed *Celtic (Kinloch Rannoch) Scottish Symphony* (1974) for the Edinburgh Festival, and in Ireland where he presented an exhibition of drawings in Dublin: *The Secret Block for a Secret Person* (1974).

Beuys's interest in materials and pilgrimage was reflected in Susan Hiller's *Homage to Beuys* which was comprised of a felt-lined cabinet of bottles of holy water from shrines and wells all over the world. She began the collection in 1969 and continued to add to it for the rest of her life, showing the work at the Lisson Gallery as *Emergency Case Homage to*

Beuys (2012) archived in thirteen wall-mounted reclaimed medicine boxes. In 2016, Hiller opened the project as a participatory call-out to other artists, saying "I would like to suggest trying the experience of being a pilgrim, instead of a tourist, by journeying somewhere to collect a special water sample" (Bore, 2016, unpaginated). Her suggestion was taken up by film-making partnership Webb Ellis on the night of the Autumn Equinox in 2016. With twenty-two others, they made a seven-mile pilgrimage to collect water from Jacob's Well in Valentines Park, Ilford. Shortly after, they made another seven-mile pilgrimage with twelve other people to Saint Hilda's Spring in Hackness, near Whitby in Yorkshire. The outcome was a fifteen-minute video, *An Empty Vesse*l (2016), which premiered simultaneously at Bow Arts in London and Crescent Arts in Scarborough (all in the UK).

The collecting and archiving of water are similarly a preoccupation of artist Amy Sharrocks for her *Museum of Water.* This collection was started in 2014 as a "live artwork" to provoke observers into reconsidering their relationship with an everyday but vital substance. The museum consists of "publicly donated water and accompanying stories" and includes "ghost water and bad dream water, water from the last ice age, a melted snowman" and a variety of spit, urine, and breath (Sharrocks, n.d., unpaginated).

Soil

Perhaps the liveliest matter of all, complex and composting, soil is composed of organic matter, minerals, liquids, gases, and microorganisms; the specific proportions of which can mean a difference between abundance and failure to thrive. Consequently, the composition of soil – the effects and affects of its active agents – also informs the artists working with it. The founder of butoh, Hijikata Tatsumi, for example, growing up in Tohuku, Northern Japan, a formerly rural region, saw "rapid and often environmentally heedless industrialization on a tremendous scale." Polluted run-off from the new factories entered the soil systems of local paddy fields resulting in "minamata disease, caused by mercury poisoning", and "itai-itai disease, which is caused by cadmium poisoning". As a result, "landscapes and communities were destroyed; traditional social relations and mores were disrupted." Within this context, Tatsumi developed a new dance form that incorporated a way of moving, with "characters who bent their knees and backs deeply with their chins sticking out" (Nanako, 2000, p. 22). This pointed to symptoms such as ataxia and bone distortion, which Tatsumi saw within the rural population of Tohuku at that time.

In 1975, choreographer Pina Bausch deployed "six dumpster-sized containers of peat on a bare stage" (Spalink, 2024: p. 42) for Tanztheater Wuppertal's 1975 production of *The Rite of Spring*. The event broke conventions as a collaborative human/nonhuman action of mutual marking; the moving bodies of the dancers stirring peat into the air and marking the thickly covered stage, while the peat "inscribes the dancer's skin" (Spalink, 2024: p.49). Performance academic Angenette Spalink suggests that while Bausch's displacement of peat from the peat bog was an extractive action, this was done knowingly. Peat bogs in their wet state are formed predominately from decaying sphagnum moss, soil, and water, and have a huge capacity to lock in carbon, but also, paradoxically to release carbon when disturbed and drained. Likewise, the peat in Bausch's *The Rite of Spring* was extracted from its environment and wafted into the air by the dancers. It was "not treated as kin," but rather as inert matter in a way that aligns with capitalist processes, and through this action, Spalink argues, Bausch "exhibits the limitations of anthropocentric world views" (Spalink, p.53). Yet, despite being treated as inert, the peat was "not stripped of its animacy" (Spalink, p.52), but demonstrated "that it is a performer in its own right" (Spalink, p.42). What is more, as "peat and humans dance together their compositions change," the mutual marking of the dance "unsettles anthropocentric boundaries" (p. 53), and the mingling of sweat and peat reminds us of the sticky porosity between entangling un/human lifeworlds.

British artist Miranda Whall takes a different approach to collaborating with soil in S*oil Voices (2024),* which was part of a wider interdisciplinary collaborative project with scientists, musicians, and dancers, *When Earth Speaks*. Rather than choosing, like Bausch, to extract soil from its habitat, Whall chose to extract her own body from its usual habitat, by lying for hours in a two-foot deep ditch she had pre-dug for herself six hundred metres above sea level on the Ffridd, a Welsh upland plateau (see *figure 5.3*). Nearby soil sensors were used to pick up fluctuating soil moisture and temperature, and this data feed was transmitted through a headset, which she relayed as a series of spoken numbers from her windswept trench every fifteen minutes. The event was filmed and live-streamed through social media. In a conversation about this work, Whall tells us how the sensors surrounded her, connecting "like rhizomes" across the ground. As well as hearing a robotic voice speaking data numbers for her to repeat, she could also hear the transmitted information as an electronic "squeaking" in her headset. After a while, this sounded to Whall like "mice or frogs" in the slatey soil beneath her. Whall acknowledges how "contradictions" arise when working in such visceral contexts while wrapped in technology; of a corporeal intimacy of being connected to the soil in such an isolated site whilst being livestreamed across the globe to colleagues, friends, and family "a few thousand miles apart" (Whall, 2025, private online conversation, dated 13.2.2025).

Whall's intention was to offer a fleshier perspective of the soil data she performed, by embedding her body "into the landscape," entangling "with the soil and its statistics to become a creature of the mud." She aimed to take a "decentred, non-hierarchical and interconnected" perspective to allow "audiences to listen, literally and metaphorically to what the earth is telling us" (Whall, in *Aber News*, 2023, unpaginated). She is also keen to

Figure 5.3 Miranda Whall in "Soil Voices: When Earth Speaks," 2024. Photo: Ashley Calvert.

emphasize that this particular performance was far from a "spiritual experience." While there were perhaps moments of contemplation or peace as she lay alone in her trench, most of her thoughts were on the organizational logistics of making it happen (Whall, n.d., unpaginated). In *Soil Voices*, Whall includes "the mountain itself, the soil, sensors, data, and robotics" as collaborators in the performance (Whall, n.d., unpaginated). The piece, intended to last twenty-four hours, was cut short at twenty minutes past four in the morning when the talkie-box data stream failed due to a broken firewire port. Although not intended, after hours of relaying data every fifteen minutes, Whall's unexpected silence had breathed something else into the work for the observer, generating unplanned readings of it, perhaps a note of caution to those relying on a technofix: "we are hanging on a thread." The firewire port's "failure" allowed space for other agencies to perform; the wind picked up and, unsure whether she was asleep or awake, those listening in were left to imagine what this fragile human body was sensing in this exposed site at night. New and ancient images floated freely: cist tombs and bog bodies were invoked, and also Schrödinger's box; Whall as his cat, equally dead and alive.

For those wishing to further explore performing with vibrant matter, there are many more examples in this mode worth tracking down, such as Peter Fischli and David Weiss's interrogation of moving matter in *The Way Things Move* (1987), Michael Landy who systematically shredded and destroyed everything he owned in *Breakdown* (2001), and, in contrast, the work of Ndaku Ya La Vie Est Belle, a collective of street performers from Kinshasa (Congo) who use street trash to adorn their own bodies, turning themselves into living and performing sculptures. Closer to our home among historical tin mines, there is eco-feminist Laura Denning's collaboration with the River Erme in Devon in *River Stories* (2018-ongoing) and Sarah Blisset and Kerry Priest's performance of scrying rituals at SOAK Live Art, casting molten tin into water in response to mining on Dartmoor in their *Alluvial Matters* (2024).

A lesson for practitioners is how riskiness, unpredictability, and lack of control so often become central when working on the scales we have discussed. Perhaps that is key: to learn to work alongside, in a kinly way with lively materials, rather than exploit, extract, command and control; whether examining the pliability of lead when you roll it, tracing toxins through bodies and ecosystems, dallying with deep pasts in ancient granite, scrying futures through molten metals, or lying in a grave you have dug for yourself on a Welsh mountain.

If, like Crab & Bee, you have been surprised to find yourself reading poems to sewage outflows, hurling logs into juddering mud and wet string at walls, or conducting rites for a murmuration of starlings, perhaps remember this: when performing with non-human actors, be prepared for the unexpected. Part of the nature of such objects is that they are always to some extent withdrawn from us – "they are merely there; they don't stick out. It's not that they don't exist at all. It's that they are less weird, less oppressively obvious versions of themselves" (Morton, 2021, p. 8) – so when in this section we discuss examples of more-than-human performances, from the hyper-scalar to the intimate, intracorporeal and elemental, we should not be surprised that all of them have pushed, expanded or exploded a usual human-centric register of perception. They offer us a thready string to reknit ourselves into "other ways to reworld, reimagine, relive and reconnect" (Haraway, 2016, p. 50) by tracing out, pointing to, and collaborating with such imperceptibles.

Gravity as a dramaturgy

Attractions

The role of gravity is crucial to diagramming. Gravity is the attraction of all mass to all other mass; it empowers the connectivity of the diagrams. It is the starting point of a diagrammatical dramaturgy. Once you make gravity the abiding principle of your performing you switch from an anthropocentric archive to an ecological repertoire.

Because humans live on a huge piece of rock and metal (the Earth), the dominant attraction is to the planet. As a result, we experience gravity as "down." The planet holds us to itself. As you walk you are constantly aware of the action of gravity; not only our own connection to the ground, but all around there is a collapsing garden shed, a landslip from a cliff, a junkyard subsiding into a valley, an incoming tide. At a cosmic scale, however, gravity is not directional, it is a promiscuous attraction of all things to all other things. When you walk into a room, the walls imperceptibly move towards you.

When making a performance, then, there is never any static "at rest" to begin from. You are always joining an existing flow of attractions. You apply your improvising and devising to an existing web of forces. In visiting any place, there is always the possibility for you to lie down, move towards, or sink in. At the same time other things are doing the same all around you. When you lift a prop, you pass on some energy to it; you have given it an energy of position, which it can spend by pressing down on your hand. Or falling to the floor. In simple actions of lifting, carrying, turning, putting down, rolling, you create a dance of different forces.

Such non-matrixed stagehand-like actions may not seem to be sophisticated. Yet, despite their apparent cultural simplicity, they are physically complex in their interplay of forces. If you begin with gravity in the performance space, as the core or base of your performance, it is more likely to be woven from ecological threads.

In *Plymouth Pantheon* we wrote of how "Dartmoor and Plymouth are two parts of the same place. Falling down is falling together" (Smoke and Mirrors, 2019, p. 64); the RAAY building where we installed our *Plymouth Labyrinth* exhibition and performed was in a state of dilapidation, subject to gravity. As we prepared a performance there, the new owners were working to keep the building up, and this fed into our work in terms of the actions of our performance/séance: raising up, picking up and putting down, pouring, scattering, and lying down. In our performance *Patterns, Diagrams and Place Stories (2024)*, gravity played a role in the movement of water through a landscape, from the top of the hills to their base. Gravity was implicit in our gathering red and white water from the springs at Glastonbury.

In our "being there" with places we often give attention to examples of human vulnerability to more-than-human forces (and *hubris* in the face of them). More conceptually and diagrammatically, however, we have drawn on the idea of gravity as active beyond a dominant planetary mass, celebrating gravity in cosmic terms as a force of connectivity; a way of conceptualizing the diagram as a set of mutual attractions. Understood in this way, the web of the diagram is not an inert set of anchoring lines but a seethe of creaturely energies that holds itself together without stasis and with which you can join.

Salvagepunk

Scavenging

It is hard to imagine that anything could be awry about the straightforward efficacy of re-using or recycling materials. Yet, various agents – companies, individuals, state entities – are not above appropriating such processes to serve their own ends (capital accumulation, profits, status, marketing, security). Indeed, as should be clear from the ethos of commercial bricolage from which the artists of Port-au-Prince's *Atis Rezistans* have emerged – "a city that seems to marvellously, imaginatively scrape by using the discards and still-ripe refuse of the overdeveloped" (Braziel, 2016, p. 419) – many disenfranchised, blacklisted and unemployed workers are by necessity forced into a shadow economy of scavenging and re-purposing.

This is not new. What is novel is how globalization has accommodated the work of such "scavengers and waste pickers," bringing them in from the margins and into "the development of global supply chains" that now support large scale industries, and high profile companies, particularly, but not exclusively, in the rapidly expanding economies of India and China (Medina, 2011, p. 1). While the recycling and "creative production of local bricoleurs" (Braziel, p. 419) are often celebrated as models of sustainability, this becomes questionable when their innovation, energy-efficiency, and poorly remunerated, dangerous labour are quietly subsumed within the supply chains of global industries that consume vast quantities of fossil fuels with resulting climate effects.

In his book *Combined And Uneven Apocalypse* (2011), the author Evan Calder Williams proposed a critical means to identify a scavenging and re-purposing of ruined and abandoned materials that resists any such subsuming and accommodation. He calls it salvagepunk; a cobbling together from post-apocalyptic fictions that refuses to celebrate their ruins but instead proposes an aesthetics of messy repairs. It comes from a seeking for "a model of *construction*" needed to get beyond the repeated dismantling – "cracking apart of the pseudo-totality of late capitalism" – by cultural and social avant-gardes that have included brutally accelerationist futurisms to norm-bending body art. Instead of repeatedly inflicting this undoing of conventions and fragmentation of cultural forms on the ever less integrated social entities left behind by repeated financial crashes since the oil crisis of the 1970s, bouts of austerity, paranoia post-9/11 and the unresolved fallout from war and pandemic, Williams proposes "the harder task of knowing when to call it a wreck and to start to dig through that wreckage" (Williams, 2011, pp. 30–31).

A sneer

Having got to this point, Williams then – and this is his innovation – swerves the habitual "avant-garde move" not by avoiding it, but by *doubling* it. At first Williams's proposals for "welding, stitching, rewiring" rather than "the smoothing of compromised synthesis," and a welcoming of "a return of the repressed idiosyncrasy of outmoded things" seem easily accommodated within avant-garde practices and familiar to performers from Karl Valentin to the "happenings" of the 1960s to the Mutoid Waste Company. However, Evan Calder Williams then introduces the crucial dynamic of doubling through a "sneer" in cyberpunk, not a sneer "at a barren speculative future" or the 1970s punk's sneer at everything, but "the hidden sneer of [the] present itself" (p. 32), something like 'the graveyard smile and the solemn gaze at comic events' of Dadaist montage (p. 37). Thus, Williams doubles and then reverses the gaze: "the 'look' of salvagepunk should be less about how it appears, from cobbled together caravans to junk-world robots, and more about a kind of look *onto* that world" (p. 37).

This looking *on* requires a distance and disconnection from supply chains, local or global; an abstraction and separation that can be used as a lens of magnification for reparable fragments. Rather than something capable of assimilation, what is rescued can then remain repressed and idiosyncratic, outside of attractive modes, uncool, without exchange value. It operates outside the frames of commercial and social capital because it is mostly constituted by its *look back* at the system from where it came – like a guerilla's or bank robber's surveillance, not to facilitate destruction or theft but transformative repair – a *retort* to its own origins, indifferent to its own identity.

In our performances, the trash and everyday objects and materials are not transformed; there is no value-added. We have rarely appeared at events with an admission charge. There is no merch, but when possible there are giveaways. At best there is an exchange of sensuousness with an audience. At worst there is a traducing of postmodern cool with our clumsy "rituals," unruly "games," resistance to immersion, and failure to upgrade our materials to the spectacular objects of an icy symbolist performance. The abject and idiosyncratic figures we have recovered from the gutters and redundant quarries – flying saucer-shaped motor parts, rusted keys, bones of an unjustifiably slaughtered deer – look *on* their origins, and *on* the lines of supply that brought them to depredation and abandonment and offer them nothing back. Following salvagepunk principles, diagrammatical performances require almost nothing in the way of new production – almost everything can be salvaged – and likewise contribute almost nothing in the way of production.

Theatrisexuality

Theatre lover

Thanks to its creaturely dimension, a diagrammatical performance can be, perhaps necessarily is, a libidinal practice. Not in the sense of presenting objects of desire to audiences or gathering sexual attention to itself by a display of eroticism, but due to its practitioners "taking performance as their lover" with consequently complex entanglements. This "love" – borrowing and adapting the notion of taking cinema as a lover from Patricia MacCormack's *cinesexuality* (MacCormack, 2016, p. 1) – rather than a preference for certain performances or performers or productions, is one of *theatrisexuality* or maybe *theatri-eco-sensuality* (see Billinghurst and Smith, 2020b); a particular relation (or set of sensual relations) with the objects and hyper-objects of performance in general. It signals a reshaping of those relations (for example, away from the psychological focus of dramas of character) that have dominated multiple forms of performance characterized by their anthropocentrism (including much that has passed for eco-theatre). This change has been characterized by Baz Kershaw as an awareness of "becoming performed by ecology" (2018, p. 43).

Patricia MacCormack unfolded the idea of *cinesexuality* by applying the "psychic topography" of Gilles Deleuze and Felix Guattari (Powell, 2007, p. 8) – particularly in relation to their ideas about becoming and the forming of self (Deleuze and Guattari, 1988, pp. 268–269) – to a reception of the extreme visual imagery of a series of baroque 1970s and 1980s genre movies. Similar to the "withdrawn" object, MacCormack characterizes cinema as something that "cannot be known," and yet "is limited only to the extent we limit desire." Assuming that the object of that *cinesexual* desire is "cinema as intensity" and invoking "spectatorship as a territory of desire," the "alliance" (characterized by extreme attention to often visceral metamorphic imagery) between a spectator and film "cannot help but transform the qualities and speeds of intensities of self" (2016, p. 1). Applied to theatre (and MacCormack explicitly says that these "intensities.... are not necessarily specific to the case of cinema"), both performer and audience member have the role of MacCormack's spectator. The thing of theatre that "cannot be known" is its material presence, its vast implied hyperobject. In other words, it is that vast "withdrawn" entity that is present, but only in part, in all its staging, props, bodies, technology, electricity, and spaces.

Larval

Although mostly taken for granted as a neutral territory, *theatrisexuality* assumes such a hyperobject to be every bit as loveable and extreme – "larval" to the extent that it is "headless and tentacled [and] unfolds and refolds" both it and us (p. 81) – as the morphing body-images of MacCormack's cinema/lover. The effect on the performance maker and audience member entering into such a relationship with performance's matters – and it is here that the relation to an ecological meshwork is explicit – is a "redistribution of self and world, self in the world, world in self, and self as world" (p. 1).

The need for an extreme relation to performance to realize the ecological possibilities of a diagrammatical aesthetics – "in the more oblique possibilities inherent in the theatrical medium" (Lavery, 2016, p. 229) – is a product of social circumstances. The social milieus that have in the past supported radical performance – the workers movement for Meyerhold, Brecht, and Piscator, the dissident underground resistance for Cricot 2 and Theatre of

13 Rows, the counter-cultural and hippie movement for the Living Theatre, or the Indian independence movement for Jana Sanskriti – are increasingly absent or fragmented. Recent progressive social insurgencies – Occupy, Arab Spring, Black Lives Matter, Extinction Rebellion – have tended to be effectively organized, open, spontaneous, and innovative, but none have been sustained. In the context of an increasingly mediatized and privatized society, *theatrisexuality* is a quietly desperate means for performance makers to develop solidarity with those (more-than-human) allies that are hyperobjective and multiple. "Quietly" because it does not need to make an exhibitionist fuss about itself; the climate crisis provides all the melodrama.

MacCormack's radical notion allows us to think about live diagrammatical performance less in terms of techniques or styles and more in terms of a shift in relations; that "redistribution of self and world." Just as she draws on Deleuze and Guattari's ideas about "becoming," so a diagrammatics of performance (through the lens of *theatri-eco-sensuality)* can work through their idea of a sorcerous pact or contagion (infatuation) with a "demon" lover; a relationship with a force that "sometimes appears at the head of the band, sometimes as the Loner on the sidelines, and sometimes as the highest Power" (1988, p. 268). This has little or nothing to do with the supernatural, but rather with those "[U]nnatural participations or nuptials [that] are the true Nature spanning the kingdoms of nature. Propagation by epidemic, by contagion, has nothing to do with filiation by heredity.... contagion, epidemic, involves terms that are entirely heterogeneous." Within such multiplicitous relations, there is little room for repetition; but what *is* repeated is a shifting between "human being, an animal, and a bacterium, a virus, a molecule, a microorganism. Or in the case of truffle, a tree, a fly, and a pig" (pp. 266–267).

Eco-aesthetics emerge

In such complex entanglements of infectious eco-attraction, conferring responsibilities but without fixed points of value, a diagrammatical eco-aesthetics of performance can emerge in the context of crisis. Not as a structure or a classification, but as a seething web across which to approach objects and hyperobjects that are in states of extreme excitation, hyper-stimulated in parts by the accelerating industrialism and information exchange powered by practices of extraction, production, distribution, and waste. With such affective tools, it might be possible to undo an understanding that has for so long taken for granted as neutral ground the harmless inertia of "props" (Smith, 2019, pp. 181–189) and the "empty space" (Brook, 1968) of the place of performance. Their flux can now be appreciated, uncaged and celebrated.

This is in tune with the emergence, most obviously (at least in the UK) in site-specific theatre, of a shifting of attention to more-than-human agents, initially in relation to places and their meanings. Academic and performance maker Mike Pearson was among those who led the way with his proposal that as a matter of principle – one he practiced, for example, in his book *In Comes I* (Pearson, 2007) – "[A]ny full description of a site-specific production surely requires as much said about the site – its architecture, its atmospheres, its climate, its history – and the emplaced scenography – its surface, substances, objects, locales – as about the themes and dramaturgy, to fully account for what happened" (Pearson, 2018, p. 300). Even more important, then, to now apply this principle concerning performance documentation to the performances themselves.

Pearson's suggestion may at first appear to be a somewhat dry one, removing human affect and replacing it with geographical details – an exemplary challenge to future performance academics, historians and author-practitioners to establish a new set of attention values and proportions – until we consider the ecstatic nature of the agencies now involved: floods, storms, wildfires, pandemics, landslides, mutations, "invasive" species of fungi, and radical behavioural changes in animals under environmental stress. Diagrammatical performances are for and about these more-than-human participants (present in materials and shapes or present by their absence in our representations/substitutions of them). Any usual final "resting place" of a performance – socially and with its audience – becomes evidently incomplete and unsatisfying; begging the question how and where our performances should now "land." Wherever that is, it is unlikely to be a place for seeking acclaim; there is no machine within these relations for social or commercial "success" (lacking the means for an accumulation of social or cultural capital). Nevertheless, just as rewilding is not about walking away – absenting humans and letting everything else "get on with it" – there is a residual attraction, a gravity of situation, that continues to draw us onto this terrain. (Importantly, we should not repeat the mistakes of the past and turn the marginal nature of diagrammatical performance into an attractive milieu; a resort to the margins is a recoiling from catastrophe not a cultivation of a style.)

The subtle but sinister return of functionalism

The *theatrisexual* renewal of performance space – both of the *theatron*, the place of viewing, and of the auditorium, the place of listening – as zones for engagements of the human *sensorium* with more-than-human desires, sets it (and us) somewhat uncomfortably at odds with a scientistic – "follow the science" – approach of much eco-culture. Performance addressing conditions of climate crisis, while drawing on standard ecological principles of sustainability and resilience, without some attention to the value of the responses of more-than-human agents, no matter the iniquity of the origins of their excesses and hyper-stimulation, can, and has, lead to a reproduction of functionalist ideas.

In *Sustainable Theatre*, Iphigenia Taxopoulou (citing Una Chaudhuri) makes the argument that "a 'useful ecological theatre' will call for a turn towards the literal, a programmatic resistance to the view of nature as commodity or its use as metaphor" (Taxopoulou, 2023, p. 157). This captures a certain problematic approach, where a "literal" receptiveness to the more-than-human (side-stepping "cultural constructs of the natural world") is upscaled and implies something like the utilitarian qualities of an instrumental application common to industrialism. Taxopoulou follows this with an assertion that "[S]eeds of such a theatre are to be found in Ibsen's *The Wild Duck*" (Taxopoulou, 2023, p. 157), which returns a troubled discourse around eco-theatre to formerly comfortable literary assumptions based on a facility to convincingly convey human arguments. The animosity to metaphor, then, when subjected to Taxopoulou's programmatical twist, threatens not just constructs of "the natural world" but – in its anxiety to find practical and resistant solutions – the desire for the theatrical aspects of performance itself.

What is missing from such an approach, despite its just attention to "the natural world," is an appreciation (let alone a celebration) of the energies released in the process of climate change, without which it is hard to generate an ecologically centred performance aesthetics. If we are now making our performances with plastic, both inside our guts and alongside us in the streets and streams, accompanied by the possibility of wildfires, floods and blooms of jellyfish, then we need an aesthetics that in some way addresses these naturally queer intensities: "[U]nnatural participations or nuptials.... spanning the kingdoms of nature. Propagation by epidemic" (Deleuze and Guattari, 1988, p. 266). Without some account of such naturally unnatural agencies, no matter how stimulated, performance (in every sense) is in danger of restricting itself to safe human solutions, fixes, programmes, usefulness, and message-theatre. It implies the valorization of an art of telling others (including a misbehaving natural world) what the right thing is to do, and what the right programme is to follow.

In conflict with this approach, and against the idea that "the natural world" is a rational agent, we embrace the idea of the composer Olivier Messiaen that "the cosmos, is made of refrains" (quoted, Deleuze and Guattari, 1988, p. 341) for which we need diagrams as means to access their patterns. Similarly, Patricia MacCormack's work on spectators' engagement with cinema provides a theorization of "becoming libidinally larval" (2016, p. 85) that does justice to the physical and imaginal intensity of performing and spectating bodies in eco-crisis. MacCormack reflects the extreme vulnerability and desirousness of humans and of more-than-humans in the places of performance. This is not quite an Artaudian model for Dionysian immersion or self-obliteration, but for a desiring space in which drama's conflicts of doing are held back or redirected, sidestepping messaging in favour of the emergence of more desires with a greater likelihood of mutation: "transformative potential is not to be found on the side of ideologemes.... the ideas and content of *what* is presented on the stage.... but rather in the modes and manners of the *how* of re-presenting: a presentation irreducible to a once and for all determinable content" (Steinhauser, 2016, pp. 25–26). Thus, the completion and resolution of human intentions are best deferred so as to be ready to respond to Hélène Cixous's refrain: "[M]y desires have invented new desires, my body knows unheard-of songs" (1976, p. 876).

Sexy theory

Talking theory

Throughout all this – walking, visiting, researching, and performance-making – we suggest always talking theory to each other. This is a practice we have described elsewhere as "sexy theory" (Billinghurst and Smith, 2020a, p. 136); the idea that sharing theoretical ideas (about anything) can and should be as desirous, tentative, tactile, and experimental with the world at large as any immersive encounter with another person. Just as we have emphasized the importance of embodied presence, we advocate for an embodied analytical mind.

Before we even began to consider performance making as part of our work together, we spent several months in a preparatory-to-the-preparatory phase (without knowing what we were preparing ourselves for), immersing ourselves in intense encounters with a series of terrains, across Southern and South West England, Wales, and the Scilly Isles. Throughout this process, we shared texts back and forwards between us – from Héléne Cixous's "The Laugh Of the Medusa" via the arachnean texts of Fernand Deligny to the "string figures" of Donna Haraway – always situated in and in trialogue with the places we were in.

Give and take

This is sometimes a process of considered rejection as much as one of critical adoption. We have resisted theorizing ideas of dwelling or belonging, avoiding an apocalyptic "re-set" or a romanticization of imaginary "returns" to a pre-industrial rural idyll or ecological "ground-zero." Instead, we have tried to give attention to the more-than-human as they are (rather than how we imagine them, or might like them to be), drawing principles and strategies from what they are doing as much, or more, than from our reading and discussing. We also pay attention to the things-to-come, and we have welcomed the arrival of new species, just as we see it as a opportunity to welcome humans seeking refuge; migrations of all kinds being part of historic global processes as well as violent disruptions. Everyone and everything – including theory – is migrant, ever since we all set out from a tiny speck of dense matter at the "big bang" on a journey of inflation.

References

Abram, D. (1997) *The spell of the sensuous*. New York: Vintage.

Adams, D. (2014) 'Joseph Beuys: Pioneer of a radical ecology'. *Art Journal*, 51(2). Available online at: https://www.researchgate.net/publication/271665093_Joseph_Beuys#fullTextFileContent. Accessed on: 3.2.25.

Bennett, J. (2001) *The enchantment of modern life*. New Jersey: Princeton University Press.
Billinghurst, H. (2019) '*Ways of making: producing artworks in the studio in response to experiential walking*'. *University of Plymouth*. Available online at: https://pearl.plymouth.ac.uk/foahb-theses-other/321/. Accessed on: 25.2.25.
Billinghurst, H. and Smith, P. (Crab & Bee). (2020a) *The pattern*. Axminster: Triarchy Press.
Billinghurst, H. and Smith, P. (2020b) 'Convivial acts for an eco-sensual labyrinth', *World Futures*, 76(5-7), pp. 337–352.
Blackmore, S. (2003) *Consciousness: An introduction*. London: Hodder & Stoughton.
Bore, C. (2016) 'Susan Hiller To Extend Her 'Homage to Joseph Beuys' With New Participatory Art Project'. *Museums EU, The European Museum's Network*. Available online at: https://museums.eu/article/details/113359/susan-hiller-to-extend-her-homage-to-joseph-beuys-with-new-participatory-art-project. Accessed on: 4.2.25.
Bradotti, R. (2013) *The posthuman*. Cambridge: Polity Press.
Braziel, J.E. (2016) 'Atis rezistans (Artists resistance): Voodoo street sculpture at the grand rue, Port-au-prince', *Callaloo. Johns Hopkins University Press*, 39(2), pp. 419–437.
Brook, P. (1968) *The empty space*. New York: Atheneum.
Burroughs, W. and Gysin, B. (1978) *The third mind*. New York: The Viking Press.
Burroughs, W. and Odier, D. (2008) *The job*. London. Penguin.
Cade, J.F.J. (1949) 'Lithium salts in the treatment of psychotic excitement'. *Medical Journal Of Australia*. 2(10). Available online at: https://pmc.ncbi.nlm.nih.gov/articles/PMC2560740/pdf/10885180.pdf. Accessed on: 21.1.25.
Callas, N. (n.d.) 'The Third Mind'. Book review. *Art Forum*. Available online at: https://www.artforum.com/columns/the-third-mind-212375/. Accessed on: 15.2.25.
Careri, F. (2002) *Walkscapes*. Barcelona: Editorial Gustavo Gili.
Carson, R. (1962) *The silent spring*. Boston (MS): Houghton Mifflin.
Chambers, D. (2022) 'The Myth of the Valley of Madness' *The Blindboy Podcast*, season 1, episode 243. Available online at: https://shows.acast.com/blindboy/episodes/the-myth-of-the-valley-of-madness. Accessed 4.2.25.
Cixous, H. (1976) 'The laugh of the Medusa'. Translated by Keith Cohen and Paula Cohen. *Signs*, 1(4). Chicago: The University of Chicago Press. Available online at: http://www.jstor.org/stable/3173239. Accessed on: 9.1.25.
Cohen, B. (2017) 'Australian Volcanic Eruption May Have Lived on in Aboriginal Stories'. *University News, University of Glasgow*. Available online at: https://www.gla.ac.uk/news/archiveofnews/2017/may/headline_523966_en.html. Accessed on: 24.1.25.
Deleuze, G. and Guattari, F. (1988) *A thousand plateaus*. Translated by Brian Massumi. London: Continuum.
Drum, K. (2013) 'Lead: America's Real Criminal Element'. *Mother Jones*. Available online at: https://www.motherjones.com/environment/2016/02/lead-exposure-gasoline-crime-increase-children-health/. Accessed on: 21.1.25.
Edensor, T. (2005) *Industrial ruins*. London: Bloomsbury.
Eliasson, Olafur. (2019) 'Behind The Scenes With Olafur Eliasson As He Re-photographs The Icelandic Glaciers That He First Photographed Twenty Years Ago, In 1999'. *The Glacier Melt Series 1999/2019*. Available online at: https://olafureliasson.net/artwork/the-glacier-melt-series-1999-2019-2019/. Accessed on: 23.1.25.
Fisher, M. (2009) *Capitalist realism*. Alresford: Zero Books.
Fisher, M. (2014) *Ghosts of my life*. Alresford: Zero Books.
Fremantle, C. (2015) 'Anne-Maria Culhane: Earthwalking'. *eco/art/scot/land*. Available at: https://ecoartscotland.net/2015/09/28/anne-marie-culhane-earthwalking/. Accessed on: 7.1.25.

Garcia, D. (2017) 'Dark jesters hiding in plain sight', in Theo Reeves-Evison and Jon K. Shaw (eds.) *Fiction as method*. London: Sternberg Press.

Hann, R. (2024) *Beyond scenography*. Abingdon & New York: Routledge.

Haraway, D. (2016) *Staying with the trouble: Making kin in the chthulucene*. Durham & London: Duke University Press.

Harding, D.E. (2013) *On having no head*. London: Sholland Trust.

Heim, W. (2005) 'Navigating voices', in Gabriella Giannachi and Nigel Stewart (eds.) *Performing nature. Explorations in ecology and the arts*. Bern: Peter Lang, pp. 199–216.

Kershaw, B. (2018) 'On projecting climate scenarios, landscaping nature, and knowing performance: On becoming performed by ecology', in Carl Lavery (ed.) *Performance and ecology: What can theatre do?* London & New York: Routledge.

Kroijer, S. and Rubow, C. (2022) 'Introduction: Enchanted ecologies and ethics of care'. *Environmental Humanities,* 14(2). Available online at: https://doi.org/10.1215/22011919-9712456. Accessed on: 3.11.24.

Lavery, C. (2009) 'Mourning walk and pedestrian performance: History, aesthetics and ethics', in Roberta Mock (ed.) *Walking, writing and performance*. Bristol & Chicago: Intellect Books.

Lavery, C. (2016) 'Introduction: Performance and ecology – What can theatre do?, *Green Letters*, 20(3), pp. 229–236.

Le Guin, U. 1986. "The carrier bag theory of fiction" in *Women of vision*, edited by Denise Du Pont. London: St Martin's Press, pp.165–170.

Le Guin, U.K. (2019) *The carrier bag theory of fiction*. London: Ignota Books.

Lowe, J. (2017) 'Lithium, Love and Losing My Mind: Jaime Lowe On Her Life With Bipolar Disorder & Drugs To Manage It'. Interview, *Democracy Now.* Available online at: https://www.democracynow.org/2017/12/28/lithium_love_and_losing_my_mind. Accessed on: 21.1.25.

MacCormack, P. (2016) *Cinesexuality*. London & New York: Routledge.

Madsen, M. (2010) *Into Eternity: a film for the future*. [DVD].

Massey, D. (2005) *For space*. London: Sage Publications.

McCraig, A. (2019) 'Lost Glacier to be Honoured with Memorial Monument'. *News and Media Relations, Rice University*. Available online at: https://news2.rice.edu/2019/07/18/lost-glacier-to-be-honored-with-memorial-monument/. Accessed on: 23.1.25.

Medina, M. (2011) 'Global Supply Lines In Chinese Industrialization'. Working Paper 2011/78. *United Nations University/World Institute for Development Economics Research*. Available online at: https://www.econstor.eu/bitstream/10419/54093/1/67836379X.pdf. Accessed on: 20.1.25.

Pearson, M. (2018) "Site-Specific Theatre". In Arnold Aronson (ed.), *The Routledge Companion to Scenography*. London & New York: Routledge, pp. 295–301.

Morton, T. (2013) *Hyperobjects: Philosophy and ecology after the end of the world*. Minneapolis: University of Minnesota Press.

Morton, T. (2021) *All art is ecological*. London: Penguin Books.

Mulvaney, A. (2019) 'Double take: the fairy bush in County Clare that moved a motorway'. *The Journal*. Available online at: https://www.thejournal.ie/fairy-bush-co-clare-4604485-Apr2019/. Accessed on: 24.2.25.

Nagle, A. (2017) *Kill all normies: Online culture wars from 4chan and Tumblr to Trump and the alt-right*. Alresford: Zero Books.

Nanako, K. (2000) 'Hijikata Tatsumi: The Words of Butoh'. *The Drama Review*, 44(1). Available online at: https://www.jstor.org/stable/1146810. Accessed on: 7.2.25.

Neimanis, A. (2017) *Bodies of water: Posthuman feminist phenomenology*. London & New York: Bloomsbury.

Noys, B. (2014) *Malign velocities: Acceleration and capitalism*. Alresford: Zero Books.

Noys, B. (2016) 'The Horror of the Real'. *International Journal of Žižek Studies*, 4(4). Available online at: http://zizekstudies.org/index.php/IJZS. Accessed on: 15.1.25.

Nun, P.D. and Reid, N.J. (2015) 'Aboriginal memories of inundation of the Australian coast dating from more than 7000 years ago'. *Australian Geographer*, 47(1). Available online at: https://www.tandfonline.com/doi/full/10.1080/00049182.2015.1077539?scroll=top&needAccess=true#d1e191. Accessed on: 24.1.25.

O'Connell, B. (2012) 'Recovering well: why do people travel from all over the country to a fresh water well in Co Kerry?'. *The Irish Times*. Available online at: https://www.irishtimes.com/news/health/recovering-well-1.556572. Accessed on: 28.1.25.

O'Sullivan, S. (2016a) 'Mythopoeisis and the Fictioning of Reality'. Available online at: https://www.academia.edu/19872052/Mythopoesis_and_the_Fictioning_of_Reality. Accessed on: 25.2.25.

O'Sullivan, S. (2016b) 'On the diagram and the practice of diagrammatics'. Available online at: https://www.academia.edu/27786648/On_the_Diagram_and_a_Practice_of_Diagrammatics. Accessed on: 19.2.25.

O'Sullivan, S. (2024a) *From magic and myth-work to care and repair*. London: Goldsmiths Press.

Oliver, B. and Whitehead, S. (n.d.) *Simon Whitehead/Barnaby Oliver. Available* online at: https://www.blogger.com/profile/01186958259010322489. Accessed on: 22.1.25.

Dyangani O. and Wa Lehulere, K. (2017) 'History will break your heart', in Henriette Gunkel, Ayesha Hameed and Simon O'Sullivan (eds.) *Futures and fictions*. London: Repeater Books.

Pearson, M. (2007) *In comes I: Performance, memory and landscape*. Exeter: University of Exeter Press.

Powell, A. (2007) *Deleuze, altered States and film*. Edinburgh: Edinburgh University Press.

Reeves-Evison, T. and Shaw, J.K. (2017) *Fiction as method*. London: Sternberg Press.

Rose, M. (2020) 'Access denied? Walking art and disabled people', in Helen Billinghurst, Claire Hind and Phil Smith (eds.) *Walking bodies*. Axminster: Triarchy Press.

Schechner, R. (1993) *The future of ritual*. London & New York: Routledge.

Sharrocks, A. (Undated) *Museum of Water.* Available online at: https://museumofwater.co.uk/. Accessed on: 4.2.25.

Shoard, M. (2002) 'Edgelands', in Jennifer Jenkins (ed.) *Remaking the landscape*. London: Profile Books.

Smith, P. (2017) 'Two Walks With Objects', *Humanities*, 6(3), p. 51. Available online at: https://www.mdpi.com/2076-0787/6/3/51. Accessed on: 13.1.25.

Smith, P. (2019) *Making site-specific theatre and performance*. London: Red Globe Press.

Smith, P. (2025) 'Three-dimensional metaphor: After site', in Victoria Hunter and Cathy Turner (eds.) *The Routledge companion to site-specific performance*. New York & London: Routledge, pp. 56–67.

Smoke & Mirrors (Helen Billinghurst and Phil Smith). (2019) *Plymouth Pantheon*. Plymouth: Crab & Bee.

Steinhauser, S. (2016) 'Of secret signals, absent masters and the trembling of the contours', in Eirini Kartsaki (ed.) *On repetition*. Bristol: Intellect, pp. 15–32.

Taxopoulou, I. (2023) *Sustainable theatre*. London: Bloomsbury.

Taylor, D. (2003) *The archive and the Repertoire: Performing cultural memory in the Americas*. Durham & London: Duke University Press.

Tsing, A.L. (2015) *The mushroom at the end of the world*. Princeton & Oxford: Princeton University Press.

Tufnell, M. and Crickmay, C. (2004) *A widening field: Journeys in body and imagination*. Alton: Dance Books Ltd.
Whall, M. (n.d.) *When Earth Speaks*. Available online at: https://www.mirandawhall.space/when-earth-speaks/. Accessed on: 8.2.2024.
Whall, M. (2023) In addition to the day job. Aber News. Available online at: https://www.mirandawhall.space/wp-content/uploads/2023/10/AU-Magazine-October-2023.pdf Accessed on: 12.1.2026.
Whall, M. (2025) Private online discussion with Helen Billinghurst, 25/2/25, Cambridge (UK).
Wilkie, F. (2014) *Performance, transport & mobility*. Basingstoke: Palgrave Macmillan.
Williams, E.C. (2011) *Combined and uneven apocalypse*. Winchester & Washington: Zero Books.
Williams, R. and Wilson, B.M.G. (2013) *Cumbrian Alchemy*. Cumbria: Unipress Cumbria.
Williams, R. and Wilson, B.M.G. (2015) 'Cumbrian Alchemy (2014-2014)', *International Conference and Debate on Radioactive Waste Management and Constructing Memory for Future Generations*. Available online at: https://inis.iaea.org/records/rzxew-qz410. Accessed on: 24.1.25.

6 Four case studies

Performing

Our creative response to our first working as Crab & Bee was not performance-based. We first wrote poems, painted and drew, wrote and decorated postcards. So, there may be a model here; to come at performance from the side, obliquely. Working around it before working on it or at it. Eventually, we fell to enacting 'rituals'; responding to the places we explored, on site, without an audience. Later, we invited others to join us to witness the places we had found and join in an action (perhaps a 'game') there. If anywhere, our performances began at the opening of *English Diagrams* (2018), a gallery exhibition of paintings and objects by Helen, during which guests were invited to participate in a 'Cabinet Meeting'(see below), setting a pattern for gatherings of multiple more-than-human agents from resonant spaces to a 'non-site'.

Since then, we have lit driftwood fires on beaches, dropped libations of Wolf Rock beer and whisky around a depredated moor, and saluted the colourless ghosts and plastic 'batwitch bin bags' caught in the trees. We have assembled a bestiary of real and heraldic animals, an orrery of found objects, and a toolkit of allies, maps, and actions for working in different forms and registers. So, we do not start from scratch with any performance. Instead, much of the work is a re-assembling of our relations with existing allies.

What has been crucial for us, when approaching anything like an ecological focus, has been to put aside any general assumptions or information we might have had about an ecosystem or ecological meshwork. We try to resist a temptation to work from a general ecological idea – even if it be an urgent one of crisis and emergency – down to (and framing) specific matters and, instead, to work up and out from particular things, and our own encounters with these things, through webs of connectivity towards something more general (in our case, a public performance).

In our attention to terrains, we have tended to skirt around dramatic acts of public trespass (although we do sometimes sneakily infringe). We have not created subversive or spectacular installations or public blockades. Indeed, we have cultivated the art of 'skulking', 'ninja'-like, of backgrounding ourselves

DOI: 10.4324/9781003606673-7

so that we are rarely noticed. Walking at most with one or two others, we have overheard what the old mines, hills, stone circles, remotest housing estates, derelict forts, council estate streams and culverts, polluted trout pools, eroding coastal paths, deforested slopes, and eccentric mausoleums have to say. We have traced the beds of missing rivers through the city. We continue to haunt the cobbled alleys of a parallel city. We gather up materials – white clay powder, rusty dustbin lids to make stencils, bagging the ashes of burned poems to scatter, collecting machine parts and gutter trash, and animal bones from fields and ditches.

We have given witness to the images and stories we collect: tales from church pamphlets and information boards, the red heart and white skull and green frog revealed in the opening of the 'book of the hill', the killing of the last wolf, an apocalyptic paint-ball gaming landscape strewn with plastic grenades and empty oil barrels. We have experienced the eeriness of a moonscape of white clay quarries, and recoiled from tiny white worms at the bottom of a holy well we had just washed in. In other words, over and above any staging processes, performance outcomes and receptions – which would be the usual priorities for most practitioners and critics of performance – in diagrammatical performance it is the place (ways in, ways out), the objects and hyperobjects, the shape of their relations, and the diagram, with its strange combination of conceptual and material properties, that are primary and ultimate.

As demonstrated in *Wookey Hole*, diagrammatical performances are ergonomically inefficient; they can be spontaneous and almost instantaneous or arise from months of preparatory walks and visits, most of which are not 'for' the performance, and often no performance results from this research. This is clearly not a commercial opportunity. Nor an efficient campaigning tool. Instead, the walks and preparations re-draw, extend, and/or re-assemble our diagram, bestiary of allies and orrery of objects and ideas. These diagrammatical performances are barely modernist; there is little in the way of disrupted narratives and the interrogation of forms for its own sake; the performances are not 'difficult', the contents – stories, actions, images – are straightforwardly expressed. With the exception of the paintings and icons, there is a definite lack of sophisticated artifice; we mostly use things to hand, repeatedly deploying the same objects and images (as a church celebrant might use the same cups, candles, and icons).

Our preparations are mostly for the sake of the doing of them, rather than any outcome. As with the walked terrains we diagram, each of our performances connects with the others in our repertoire; there are no 'stand-alone' projects. We have experienced the making of the performances as powerful catalysts of personal change, but we never know what change they will bring. Similar, in terms of proportions of energy, to the priorities of the actors of Jerzy Grotowski's *Teatr Laboratorium* in the 1960s and 1970s, preparation and development of meaning is always vying with, and sometimes overcoming, the popular reach of performance. In our case, the daily training of an actor-monk has been replaced by our almost daily walking and visiting of places, acquiring tacit knowledge, gained implicitly through ambulatory action, to be (possibly, hopefully) re-iterated through the performances. When there is a performance product – as with Matthew Barney's three-part process to creativity, of Situation, Condition, Production (Spector, 2004, pp. 5–6) – it only appears after many repetitions of and returns to the earlier parts of the entire process.

Performance 1: *Wookey Hole*

In this section, we show how a wide historical and geographical context informed just the small *Wookey Hole* part of *Patterns, Diagrams and Place-Stories* (2024), a diagrammatic performance made for the 4th World Congress of Psychogeography in Canterbury (UK). This is done by mapping six walks, taken between 2019 and 2023. Each walk, explored at the time as a granular "field" through reflective, written accounts and desk-based research, overlaps and intersects with other "fields" to show how the accumulating insights, stories, and emotional residue gained through repeated walking were amalgamated and applied to performance. While we do not explicitly discuss the hinterlands of the following performances discussed in this chapter in such depth, it may be understood that a similar volume of acquired information is collapsed into each part of every performance we make.

Walk 1: Priddy to Wookey Hole

We have walked from the tops of the East Mendips on our way to Wells. Descending the steep road, we are treated to views of the Somerset levels on our right, and the approach to Ebbor Gorge on our left. The road flattens out at the bottom of the hill, and we are surprised and amused to see a large, creepy plastic gnome in the road. He wears a bright blue police uniform and carries a sign: "Caution Kids Crossing." A little further on and we peer down over the wall at the enclosed yard of an old paper mill. We can see a cave entrance and a sizable river flowing out of a hole in the rock. In the yard, we see a "pixie" (an employee in costume) sweeping up around King Kong and a miscellaneous selection of enormous, brightly painted "life-sized" dinosaurs plus a mammoth (apparently, historical accuracy is not important here). Reaching the village, we peer wistfully through the windows of the "Tudor-style" Wookey Hole Inn at the seductive comfort of the cozy lounge inside. There is a lot here; it feels like a puzzle, and we agree that this is a place to explore further.

Our first foray – by chance – to Wookey Hole (a series of limestone caverns in Somerset, UK) was on the last of three days walking from Bristol to Wells in June 2019, as research for our book *The Pattern* (2020). Prior to its present-day manifestation as a tourist entertainment attraction, Wookey Hole was a significant industrial site. Powered by the waters of the River Axe, paper mills have been operated at Wookey Hole since 1610, and the milling of corn there is mentioned in the Domesday Book of 1086.

In the early twentieth century, the caves were owned by the Hodgkinson family as part of the estate of Wookey Hole Paper Mill. George Adlam – a dairy farmer whose fields were once where the Wookey car park stands today – supplemented his income by taking tour parties into the caves for the price of one shilling a head. Wookey Hole Caverns, as they were then referred to, were opened officially to the public as a visitor attraction in 1927. Subsequently, the caves, which maintain a steady 11 degrees Centigrade all year round, have been used to keep and mature cheddar cheese, as storage for the waxworks of Madame Tussauds, and have been a popular (if sometimes fatal) venue for cave divers (Southwest Heritage Trust, n.d.).

Today, Wookey Hole Caves and Attractions are owned by the Cottles (Wookey Hole Caves, 2024), a famous circus family, and welcome approximately 250,000 visitors each year to tour the eight available underground chambers (there are more). Other attractions on the site include the old paper mill, a circus school with theatre, a cinema, a penny arcade, a hall of mirrors, a (cold and damp) "play barn," a museum of diving, and a resident witch. You can also stay in a hotel with a turret on the roof that is shaped like a witch's conical hat.

Walk 2: Ebbor Gorge

After peering gingerly a long way down into the gorge from the top of the cliff (we both suffer from vertigo), we set off down a sloping woodland path. Quite a long time later (feeling like 20 to 30 minutes), we reach the bottom of the valley at the start of a tapering ravine. We follow the path into the narrowing gorge. The ravine becomes so narrow we can touch both sides of its cliff walls at once; here the path rises steeply while winding further into the crevice. We scramble up and into the rocky fissure and very quickly re-emerge into a woodland clearing. We are both astonished to find that we are back where we started – almost as though we have floated up the steep rocky chimney – at the top of the cliff; some crumbs on the path from our snacking on oatcakes earlier are evidence that we are at the exact same place. Once again, we set off down the same sloping path towards the bottom of the valley, this time walking in silence. Time, space, and gravity are playing tricks on us: up is down. The journey back down to the floor of the Gorge seems to stretch, taking even longer than the first descent.

In January 2020, we returned to Somerset for a walking weekend, to explore Ebbor Gorge more thoroughly. The Mendips, of which the Gorge is a part, are mainly formed from sedimentary rocks, largely Carboniferous Limestone ranging in age from Late Devonian to Mid Jurassic. These sedimentary rocks were formed by the accretion of debris, such as sand, mud, and organic matter, gathering on ocean floors and riverbeds, which was then compressed and hardened into rock over time. The rocks forming the Mendips have "often been uplifted, folded and faulted into complex geological structures, or eroded to produce unconformities." The karst of the Mendips was created by the "action of water dissolving and eroding limestone creating a landscape in which much of the drainage is underground." This includes "dry valleys and gorges, sinkholes and sinking streams" (British Geological Survey, 2024, unpaginated).

Ebbor Gorge, formed during cold phases of the Pleistocene by summer meltwater run-off, is a part of a National Nature Reserve located on the Southerly descent from the tops of the East Mendips towards the Cathedral City of Wells. The water that carved out Ebbor Gorge eventually etched its way into and under the limestone karst of the Mendip Hills, forming the underground river that now flows out of Wookey Hole as the River Axe.

Walk 3: Gulley Cave

We stay in the Wookey Hole Inn; the windows of which we had pressed our noses against two years before. Outside the weather is terrible, cold with heavy rain; inside, it is warm and cozy. Nevertheless, we are aware of the looming presence of the limestone karst of Wookey and Ebbor above us in the North; hot water thunders out of hotel bath taps with tremendous force, and there is no phone signal. It rains all night.

The following morning, we walk up towards the cloud that rests along the ridge above us, to our rendezvous with paleo-archaeologist Danielle at Ebbor Gorge car park. We huddle together to see part of a trove of objects excavated from a nearby cave over more than ten years. From the boot of her car, Danielle holds up, then passes around, part of a bison foreleg, nearly 40,000 years old, it is comically large like a cartoon bone. We are in awe.

She conjures more ancient bones and teeth from the cavernous boot of her car; bear, hyena, horse, wild cat. Danielle's knowledge is vast and deep. She and the cave are entwined; she tells us that this is her life's work. She dreams that one day her ashes will be scattered here. She tells us how she lies awake at night, mapping the deep-life-time of the cave; ice flows breath in, ice flows breath out; ecologies and terrains, plants, animals, rivers, and rocks are all arranged and then rearranged.

We descend through the layers of information held within Danielle's body; excavating time, marking off radiocarbon dates, sifting through strata of substrate, sandstone, and limestone until eventually, out of the boot of the car, comes the only evidence of human activity in Gulley Cave: a single tiny flake of napped white flint.

Bones firmly re-entombed in the boot, Danielle leads us to the top of the Gorge, slipping on wet leaves as we go. We recognize the location from two years before, when time, space, and gravity played tricks on us. As Danielle takes a left-hand turn towards her cave, we realize that on our previous visit we were unknowingly circling around Gulley Cave. We slither warily a short way along the path, trying not to look down into the drop below. At last, we stand as a group on a platform in front of Gulley Cave; a compact space, secured by an iron door, with a hole in the rock ceiling shaped like a two-dimensional map of Africa.

Inside the tiny cave, Danielle tells stories of community and human endeavor; the intimacy of archaeological teamwork in a confined space, the heat of summer sun bearing down into the cave every afternoon. She speaks of arduous, physical labour; buckets of sieved stone and bone carried all the way – down then up – to the car park (where the eccentric Bob lived for a while in his car) at the end of each day. She remembers her relief when a local farmer granted permission to drive onto the nearest field so the archaeologists could simply haul their load over the fence every evening. She recounts how the annual summer digs are funded by contributions from the local history society, how local volunteers bring bowls of food to sustain the archaeologists.

We bid farewell to Danielle who heads back towards her portable museum in the carpark, and we make our slow descent down the hill back towards Wookey.

Mendip limestone,
the hardest of cheese, riddled with holes
broke Bob's teeth
and bears' bones,
fang of wild cat
and hyena
crunched time
compressed space.
A mason jar, facing West;
perfectly preserved,
potted
and hermetically sealed.
Disrupting signals,
looping futures,
proffering pasts,
Big Time in a small space like the boot of a car.
A grave
full of life
unsettling gravity.

In November 2021, we were invited back to Ebbor Gorge, for a weekend field trip with the research group Making Routes, to visit Gulley Cave in Ebbor Gorge with palaeontologist and climate scientist, Danielle Schreve. As Professor of Quaternary Science in the Department of Geography at Royal Holloway, University of London, Danielle (now Heather Corrie Chair in Environmental Change at the University of Bristol) has led the ongoing excavation of this site of palaeological significance since 2006. Its organic deposits range from 11,000 to 40,000 years ago, and the study has produced valuable insights into biological responses to periods of abrupt climate change.

Walk 4: Wookey Hole Caves

It is the second day of our Making Routes research trip. We take a tour of Wookey Hole Caves with the rest of the team. It is approaching Christmas; the caves are bedecked with 1970s decorations. A devilish head of Saint Nick (or is it Old Nick?) leers over the entrance to the cave. Piped carols jingle tinnily over the rushing sound of the underground river as we wind around purple, red, and blue spot-lit pools, down shadowy passageways via vaulted chambers. The atmosphere is intense, almost nauseating; an affect induced not just by the dripping walls and gloom of the caves (which are impressive) but by the volume of aging twentieth-century tourist tat and other psychic detritus accumulated within the caverns. Our tour guide, a Christmas elf, leads us to "The Witch's Kitchen" (Chamber 1). He points a torch at a stalagmite and tells us that this is the Witch of Wookey, turned to stone by holy water. Nearby is "The Witch's Dog." We are invited to view "The Witch's Chimney" in Chamber 3.

In 2007, a large collection of inscribed ritual protection marks – also known as "witch's marks" – was identified on the rocky shaft called "The Witch's Chimney" in Chamber 3 of Wookey Hole. The protection marks are believed to date from between the sixteenth and eighteenth centuries, during which there was a "continued and widespread belief in witchcraft" (Binding and Wilson, 2010, p. 48). "The Witch's Chimney," a "closed aven" can create a "convection draught caused by the body heat of those standing below it"; spelaeologists Binding and Wilson suggest that "[I]t is reasonable to assume that early observers experiencing the appearance of chill draught may have believed this to indicate the presence of some form of evil spirit" (p. 48).

Visitors were not new to Wookey; excavations of one chamber have revealed a "Romano-British" graveyard, now flooded (Hawkes, Rogers, and Tratman, 1976, pp. 23–52). Around 1470, William of Wocester describes seeing "the figure of a woman" (but not yet seen as a stone witch) at Wookey (Binding and Wilson, 2010, p. 60). While not yet tied to witchery, it appears to have been considered a place of power, possibly embodying or facilitating oracular forces, for Camden's *Britannia*, published in 1610, states:

> Among these hils there is a cave or denne farre within the ground: wherein are to be seene certaine pits and rivelets, the place they call Ochie-hole, whereof the Inhabitants feine no fewer tales, nor devise less dotages, than the Italians did of their Sybils cave in the mountaine Appeninus
>
> (quoted, Binding and Wilson, 2010, p. 60)

In 1628, lawyer Bulstrode Whitelock described a candle-lit guided tour of the caves to see various named candle-lit rock formations, including "the Porter, the Witch of Ochies Hole, stones resembling their names." It appears that from the mid-seventeenth Century onwards, the tale of a malevolent witch and her confrontation with a minister of the church began to appear in English poetry concerning the caves (Binding and Wilson, 2010, p. 60).

Sylvia Federici argues that the shift during the seventeenth century from "a subsistence to a money-economy, in every phase of capitalist development" (2004, p.74) was inextricably entangled with increasing control of women's bodies and the persecutions of women as "witches." We can detect such forces at work within the micro-economy of the Wookey community by tracking the change in stories told about Wookey Hole Caves – from a place of memorial and oracular power to a place of evil – alongside early entrepreneurial moves to develop Wookey as a tourist attraction. Additionally, the cultivation of sensational stories about "the curse of the Witch of Wookey" was useful not just to attract wealthy day-trippers to the site with the promise of a vicarious thrill, but also as a scapegoat deflecting attention from industrial pollution (see below).

Walk 5: Warren Farm to Velvet Bottom

From the top of the hill at Warren Farm you can see the Bristol Channel. Descending into the dry valley below, it is obvious that the ground was extensively rearranged a long time ago for some purpose, but hard to read. At the bottom of the valley, I chat with a man who is photographing mushrooms and he tells me this site was once a Roman quarry.

In September 2022, Helen attended a wedding at Warren Farm, a glamping site near the village of Charterhouse on the top of the West Mendips. To clear her head after the wedding toasts, she took a stroll down into a mossy valley, wandering a little further than she had intended. Velvet Bottom and Ubley Warren are the remains of a complex of ancient Roman lead works (UK SouthWest, n.d.). Now a nature reserve and part of the "Cheddar Complex Site of Special Scientific Interest," this lead-contaminated "gruffy ground" is the habitat of several scarce, lead-tolerant plants including Dwarf Mouse-ear, Spring Sandwort, Autumn Gentian and Early Purple Bee and Frog Orchids (Somerset Wildlife Trust, n.d.).

Walk 6: Velvet Bottom to St Cuthbert's Lead Mine

We walk Eastward along the dry valley of Velvet Bottom, passing through a rising series of rectangular pits and eventually to the bumps and rough grass of Ubley Warren. At the top of the valley, close to the road to Charterhouse, we find a mass of shiny black rocks scattered over the ground; spoil from some smelting process. We see a fat, three-foot-long adder coiled on a rock. So well camouflaged is it by its green and cream pixilated markings that the snake appears to shift in and out of view. It flicks its forked tongue and basks on the hot rock in the bracken.

The nearby St Cuthbert's Lead Mine was, like that at Ubley, first opened by Romans. Subsequently reopened in the fourteenth century, it was worked intermittently until 1908. Close to St Cuthbert's Lead Mine is the sink-hole entrance to St Cuthbert's Swallet, where three of the tributary streams of the River Axe enter the limestone karst "after flowing through disturbed lands" of a thousand years of mining (Stenner, 1978, p. 49). Sample testing of Saint Cuthbert's Swallet and Wookey Hole Cave sediments has shown conclusively that "the higher levels of lead" were "un-natural" and "the result of pollution by the lead industry" and that "the lead which contaminates the Wookey Hole sediments was transported via St. Cuthbert's Swallet" (pp. 50-51). Perhaps it is significant then, that in 1681, John Beaumont described how the deaths of cattle feeding on pastures flooded by the Axe were attributed to lead poisoning (Binding and Wilson, 2010, p. 61). Binding and Wilson suggest:

> During such deeply suspicious times, when women were often accused of bewitching both livestock and people causing them to both sicken and die, phenomena such as the one reported by Beaumont could well have been taken as evidence of the association of the cave with the various forces of evil that were believed to play a large part in visiting trials and tribulations on the world.
>
> (2010, p. 61)

Performing *Wookey*

Crab & Bee perform *Wookey Hole* as a section of *Patterns, Diagrams and Place-Stories* for the 2024 4th World Congress of Psychogeography in St Paul's church hall near the centre of the medieval city of Canterbury in the South East of England. It is a relaxed affair; the audience sit around tables drinking cups of tea.

1 **Geographic context**
   The performance begins with an anecdote set in Canterbury, and from this starting point, Crab & Bee use ribbons to map out a story-terrain across the floor of the performance space, travelling North from Canterbury up Watling Street (a red ribbon) and then West along the chalk Ridgeway (a white Ribbon). *En route*, they stop to tell place-stories at key points; Royston Cave, Dunstable, Avebury, until they reach Llaneirwg (St Mellons, Cardiff, Wales). Here Phil tells the story of Crow's flight across Somerset towards White Hill in London, carrying the head of Bran the Blessed. This trajectory is mapped out with a black ribbon.
2 **The Wookey complex embodied**
   When Crow (Phil) rests at Wookey Hole, the black ribbon is arranged into a circle and Helen, wearing a black academic gown, steps into the black circle facing the audience. She dangles a plumbline – a lead fishing weight on a length of twine – from her left hand.

Here, Helen begins by telling the "traditional" story of the witch of Wookey Hole: how a once beautiful but now bitter, hateful old woman who lived in the cave was blamed for casting a terrible curse that brought blight to the land. She then unpicks this story by telling of the historic lead mining on top of the Mendip Hills; how for centuries lead particles have been washed into St Cuthbert's Swallet where rainwater enters the limestone karst, gushing downward under Ebbor Gorge and eventually out of Wookey Hole as the River Axe. She reveals how the "symptoms" of the witch's curse – cattle dying, infants failing to thrive, delusional paranoia, superstition – are symptoms of lead poisoning. Thus, she declares, the historic, systemic lead poisoning of the Wookey complex *is* the curse of the Witch of Wookey.

As Helen stands in the black circle to tell this story, she diagrams the Wookey complex with her body. She indicates Saint Cuthberts lead mine atop the Mendip Hills by gesturing to the left of her head. The plumbline (from the Latin *plumbum* meaning lead) dangling from her left hand marks out the vertical descent of the lead, the force of gravity, the downwards vector of water, and the depth of the cave. Helen indicates the winding underground flow of water using snaky, downwards winding gestures with her right hand, and then crouches and uses both hands to indicate the issuing of the River Axe out of the cave (the black ribbon circle) at her feet.

The *Wookey* diagram

The diagram that emerged from repeatedly walking the descending terrain between Ubley Warren and Wookey Hole was like a multi-dimensional jigsaw puzzle; each walk, each "field" of discovery interlocking and intersecting with another. Through our repeated walkings, weavings, and supporting desk-based research, certain corporeal and material forces working at a variety of spatial and temporal registers within the Wookey complex were revealed to us. While this set of relations resonated with those diagrammed in the "official" legend of the Witch of Wookey, also revealed were implicit structures of power and control held within this version of the story. The legend of the Witch of Wookey holds connotations of colonialism, capitalist extraction, institutional misogyny, and sexist scapegoating.

It was during the writing of a re-assembled version of the Wookey legend for *Crab & Bee's Matter of Britain* (Peakrill Press, 2025), that the accumulated insights we had assembled from our repeated visits and performance were fully gathered. The re-telling of this story helped us to more fully grasp the implications and interconnections of what we had found here: that the legend of the Witch of Wookey describes a set of relations, from the historic lead pollution of St Cuthbert's Swallet to the continuing ecological and social impact of the poisoning of the River Axe. It is the understanding of these relations – along with a critique of the diagram of power contained within this legend – that was subsequently set to work for our performance in Canterbury.

Along with her formidable, explicit knowledge of her subject, Danielle Schreve's implicit understanding of Ebbor Cave, gleaned from hours spent sifting through its stoney layers, was available to us when she talked about the place. In a comparable way, Crab & Bee were able to use a complex hinterland of embodied walking experience, re-iterated through the Wookey performance in verbal re-telling and critique, but also in hand signals and embodied gestures, in collaboration with a black ribbon, a lead weight, a piece of string, and the force of gravity.

Storming Sky and Jamie Mcphee propose that to the animist mind of the pre-Roman British, "Taranis was not a transcendent god of thunder [....] That was the Roman translation (*interpretatio romana*). Taranis was thunder" (2015, p.224). They draw from the ideas of archaeologist Miranda Green, who similarly suggests that "Sequana was the River Seine at its spring source; Sulis was the hot spring at Bath, not simply its guardian or possessor" (1997, p. 24). So, the Legend of the "Witch of Wookey" is not a representation of a fictional character but carries (uneasy) ecological relations between human and unhuman. The curse of the "Witch of Wookey" *is* the poisoning of the River Axe.

Performance 2: *Leadworking*

Crab & Bee's 2023 *Leadworking* addressed the mining out of the land at Hemerdon and Lee Moor, on the edge of Dartmoor (just beyond the bounds of the National Park), for the extraction of tungsten and white clay, as well as the historic extermination of wolves and red deer on the moor. Along with our previous performance, *A Séance For Stonehouse* (also 2023), it represented a qualitative shift from the loosely cartographic diagramming of *Nets, Webs & Carrier Bag* and *Around The Hills* (both 2021) to an attention to place in which materials, icons, and stories had greater autonomy in performance.

Indeed, with *Leadworking* it is possible to diagram the performance, describing it across seven layers of action.

1. **Rejigging the space:** *Leadworking* was performed in the attic space of Leadworks, a former paint factory in Stonehouse, Plymouth, as part of a SOAK evening of live art organized by Sarah Blissett and Kerry Priest. The performances prior to ours were performed from an "end-on" space with the audience seated in a conventional "proscenium" style arrangement. We had the seating (chairs) removed and the audience sat or stood in a semi-circle facing a side wall. On this lengthy wall were two hip-height ledges; on one we placed an icon-like painting of a wolf-headed saint, and on the other a faux-heraldic "shield" with "spear" (a found iron panel, battered and rusted, roughly the shape of Britain, and a metal railing).
2. **Establishing a convention of flexible patterning:** we unrolled a patterned Persian-style carpet and emptied onto it a number of stones, some

pebble-like, some larger (about the size of a plum). We then played a game. Taking turns, we moved the stones about, sometimes "taking" a stone or stones (as in chequers) and clearing the notional "board" (the carpet) or making a move that required the rearrangement of all the stones and the restoration to the "board" of those stones "taken." We were improvising the "rules"; once it felt as if a logic was emerging, we would change the "rules" and move the stones differently. We were encouraging the audience to see a pattern, but always ready to shift to a new one.

3 **Alchemy**: clearing the stones away, we carried in an old wooden chest, emptied it of its contents (lead sheets, rusted key, white clay powder, other materials from the mine and quarries, and gold jewellery). Turning the chest upside down, we used its base to lay out the materials in an alchemically inspired pattern; lead at one end, gold at the other. We were fixing various materials in place, while implying, by their pseudo-magical arrangement, fluidity at the invisible level.

4 **Wedding**: we asked for a volunteer from the audience, and "married" them (and, by association, the audience) to the wolves ("a pact with the pack"). This ceremony included sipping "Legend" beer and the binding of wrists with red ribbon referencing the hunting lines of the wolves. (This "wedding" was modelled on a previous improvisation in which we had married ourselves to the wolves in a wood on the edge of Wolf Minerals' Hemerdon tungsten mine.)

5 **The tangent**: we then ran a red ribbon from the wooden chest, through the audience, and towards the North (although we had dropped the cartographical-diagrammatical element, we had still oriented the performance to the actual four quarters beyond the performance space). Helen, her head draped in a red cloth (fairy tale reference explicit) then carried lead sheets and sheepskin along this ribbon, that was (to us) ambiguously both hunting line and trade route, stalked by Phil as "Big Bad Wolf." Helen stopped, turned and looked behind her every few steps, and when she did this, Big Bad Wolf froze (referencing the children's games of Granny's Footsteps and "What's the time, Mr Wolf?" and the traditional panto refrain "He's behind you!") Helen took the iron railing and killed the wolf, skinning it (removing the overcoat from Phil) and dressing herself in its "skin" with the red "riding hood" cloth now draped over her arm.

6 **Weird icon**: Helen in the wolf-overcoat stood at the head of the altar-chest; Phil placed the painting of the wolf-headed saint (a reference to Saint Hubert, patron saint of werewolves) in the crook of one of her arms, while she held the spear-railing in the other. With a piece of red ribbon Phil tied the large, rusted key (found in the tungsten mine and kept as a reference to Saint Hubert's key, once used to cure lycanthropy and rabies) to Helen's forehead. Phil took green ribbons attached to the insides of the pockets of the overcoat now worn by Helen and threw them over the roof beams of the building.

7 **Invitation:** we distributed hand-drawn maps to the audience, encouraging them to visit the tungsten mine and white clay quarries on which the performance was based.

Preparation

Leadworking emerged from over ten years of visiting the tungsten and white clay mines North-West of Plymouth for Helen, and five years for Phil. These visits included a number of striking experiences: the disappearance of an entire hill to mining, hearing the howling of wolves (that turned out to be a pack of wolves kept in the nearby Dartmoor Zoo), finding an apocalyptic paintball zone with a landscape of empty oil barrels, harassment from security guards, the ever-expanding lunar landscape of the white clay pits, the collapse of the mining company Wolf Minerals, the flooding of trout pools with white clay, a small herd of fallow deer racing across a landscape of churned mud, and an emerging alternative biome among spontaneously rewilding former quarries.

Hemerdon mine is an abject terrain – to which almost no one goes; we have very occasionally met a walker and once walked into a strange encounter that was possibly a drug deal or a sexual transaction. Despite its obvious visibility from parts of the city of Plymouth, the inhabitants of the city seem to know almost nothing about what happens in this place.

There, we found – and celebrated in *Leadworking* – certain energies. These included hunting by the now extinct wolves and the agencies of metals and clay undone from the ground. Through our repeated returns to and work with this potent site, we developed a complex lexicon of colours and materials which we deployed but never explicitly stated in the performance. It was important that this lexicon was not a symbolic system, but a way for us to locate ourselves within the vast mining complex, with its red tungsten soil, white clay, green frogs, black crows, and white bones scattered liberally across the ground. We took our cue here from Tim Ingold who suggests "images do not stand for things but rather help you find them" (2010, p. 16). The lexicon connected us to key points just beyond the site, such as Saint Hubert's stained-glass window in nearby Harford Church, the close-by pack of wolves enclosed in the zoo at Sparkwell, and the Tory Brook flowing from mine to sea, often running red with tungsten wash off or white with clay. The lexicon also helped us to locate ourselves within a cat's cradle of our previous work and collaborations with others.

We are painfully aware that the mining at Hemerdon is a continuation of the extraction of tin, arsenic, copper, and silver that not only empowered the Bronze Age civilizations of the Middle East but later financed British and other imperialisms. There is a certain melancholy in *Leadworking* which is part of a colonial political economy of affect that is oppressively powerful even when it is not evident. In the performance we protest the eradication of species and the trashing of the land (including the undermining of numerous sites of Bronze Age veneration), but we also attempt – through a faux-alchemy – to remake those abject circumstances by melding them with the energies of the emerging metals and clay and of the re-emerging species. In recognition of this mutant landscape, we proposed a new iconography of hybrid wolf-saint (melding the ghost wolves of the past with the imagery of a local saint, Hubert, the saint of hunting, woodsmen, eco-warriors and werewolves). This was not a utopian or revolutionary image, but a remaking of something salvaged from the ruins of the present.

Performance 3: *Scoriton mysteries*

The site

For this performance we were attending to the site of a 1965 "alien encounter" experienced by a care home handyman, Arthur Bryant, close to the village of Scoriton in Devon (UK) on the edge of Dartmoor. Materials were key to this performance, such as the bones of a deer found on the moor, a collection of UFO-shaped "trash," and a (first edition) copy of the book written in 1967 about Bryant's "encounter" by ufologist Eileen Buckle. We also repeated certain actions, including ritual gestures made by members of the Aetherius Society (a "ufo religion") performed on Dartmoor in 1965 according (possibly erroneously) to Ruth St Ledger's *The Witchcraft and Folklore of Dartmoor* (1972), and Phil's spontaneous response to a "pixie ring" during a field visit to Scoriton Down, in which he unknowingly echoed an Aetherius gesture.

Patterns were also key to the performance, building on the looping and repetitions we had found effective in our *A Séance For Stonehouse* (2023). For a while these were a little too effective, as we confused ourselves in the loops, and so removed some of our circumambulations in favour of moving objects. Rather than being predictable, we found that repetitions (when subject to loopy, non-linear time) invited difference, "generative of new ways.... a dynamic repetition, evolving through time" (Kartsaki, 2016, p. 4).

In *Scoriton Mysteries* we drew attention to the circular shape common to both the Bronze Age stone circles on Dartmoor and to the flying saucers as described by contactee Arthur Bryant among others. We repeatedly made circles of bones, stones, and ribbon, or carried objects in a circular motion. The audience were asked to form a circle around a space defined by four "stations" oriented to the cardinal directions. We used a circular saucer-shaped bowl during the performance and, later, a round dustbin lid. We ended the performance by "closing down the circle" – gathering up a green ribbon in a clockwise direction – in the way a procedure in ceremonial magic might be "closed down."

Despite their promise to Arthur Bryant in 1965 of the benefits of superior technology – "We are going to bring proof of a wonderful existence, and of life beyond your understanding" (Buckle, 1967, p. 63) – the aliens and their craft were represented in the performance by trash. By saucer or rocket-shaped pieces of plastic and metal rubbish we had gathered up over some years from gutters and waste grounds, or by items collected from junk shops and the local Scrapstore. We drew upon the ways that tired notions of UFOs and aliens (with attendant paranoia around government cover-up and abductions) are repeatedly recycled. We applied Timothy Morton's argument – that there is no "away" for a disposal of things (2022, p. 46) - to cultural information (whether about the mysterious purposes of stone circles or of UFOs), and applied Evan Calder Williams' notion of "salvagepunk" as the "vision of a broken and dead world, strewn with both the dream residues and real junk of the world.... shot through with the hard work of salvaging, repurposing, détourning, and scrapping" (2011, pp. 19–20). We repeated the claims of contactee Arthur Bryant and the rituals of the Aetherius Society without parody, only pointing out the saucer-shaped hills adjacent to Bryant's encounters to highlight a recurrence of shapes.

After the performance no one wanted to talk about UFOs, save for one audience member who had witnessed a satellite over the moor the night before. Instead, many were intrigued by the materials, in particular the green ribbon used to join the audience into a circle (evoking a mycorrhizal grass "pixie ring") and the black treacle "fuel" we had poured into the white saucer-shaped ceramic bowl (an oblique reference to the coffee/oil energy powering David Lynch and Mark Frost's *Twin Peaks* mythos). An email received from an audience member the day after the performance referred to "[T]he lovely circular winding shape of it.... lingering in me.... please know I.... enjoyed the shape of it unfold and unravel" (Mason, private email, received 1.2.25).

Devising preliminaries

In the early stages of the *Scoriton Mysteries* project, we focused on visiting and revisiting the "site," looking for traces of place-stories, working with fragments of uncertain experience (our own and those reported by other unreliable narrators), while aware of a general lack of agreement around interpretations of both flying saucers and Dartmoor. We were also sensitive to operations of time; how modern depredations to the terrain may echo longstanding ruptures of human/more-than-human relationships, and how experiences of looping time seem to occur in certain places and generate significance there; "missing time" being a notorious feature of UFO encounters.

Choosing a narrative of alien encounter reflected our shared suspicion that the ambiguities of the UFO phenomenon and the failures to explain it (or explain it away) say something important, if cryptic, about hyper-modernity. We chose Dartmoor as a way of addressing human despoliations of a more-than-human terrain that are as ancient as they are modern. Our reason for focusing on the stone revenant of rows and circles was how they might be an expression of attention to an ancient moment of disconnection from an immersion in everything other. This certainly reflected our experiences out on the moor, as a place of disconnection; and informed our apprehension of what these formations might mean now. Particularly – and rather banally – an affordance of a circular shape appearing in the adjacent narratives of the UFO encounter and the Bronze Age revenant.

A "dragging in" of subjectivities may threaten to undermine a project committed to attending to the active agents of a place, but we have found it also helps to disturb ourselves from the position of competent organizers in control of our materials. This has helped us to avoid performing our own expertise at the expense of more-than-human others and puts our subjectivities (not to speak of our objectivity) in question within our own project. We are familiar with uncertainty and disturbance accompanying our making of diagrammatical performances, but the slipperiness of our subject matters was particularly intense both in the development of *Scoriton Mysteries* and in writing this section.

The shape

The beginning of the modern "UFO phenomenon" is usually dated to 1947, with the sighting by a private pilot, Kenneth Arnold, of a group of nine objects flying at great speed and in formation over the state of Washington, USA. Newspapers and broadcast media reported that Arnold said that the objects were shaped like saucers. Despite these reports, widely broadcast across world media, Arnold always insisted that he only ever described the objects' *motion* as like saucers skipping across water and never called the shape of the objects either saucer or disc-like (Sagan, 1995, p. 70).

Nevertheless, for many decades, "flying saucer" was almost synonymous with "UFO." A novel by Bernard Newman entitled *The Flying Saucer* was published as early as 1948 (a tale of multiple governments conspiring to create a UFO hoax!) The term and the saucer shape of craft were repeatedly used in movies like *The Flying Saucer* (1950) and *Earth Versus The Flying Saucers* (1956), and in popular books like Donald Keyhoe's 1950 book *The Flying Saucers Are Real* and *Flying Saucers Have Landed* by George Adamski and Desmond Leslie (1953). Despite many subsequent UFO witnesses reporting triangles, orbs, bells, acorns, and cigar-like or cylindrical craft, or lights rather than structures, the shape of a solid "saucer" remained remarkably resilient, despite (or maybe because of) its lacking any authentic original. Which begs the question as to whether the saucer shape itself was independently culturally charged (Jung, 2002, p. 15), and its association with aliens from other galaxies fortuitous, but not fundamental. Or whether "saucer" operates more like a meme (Blackmore, 1999) than a description; a sign piggy-backing on the experiential intensity of the UFO phenomenon in the interests of its own dispersal and multiplication?

Either way, these instabilities echo a similar uncertainty around Bronze Age stone circles. Where identified, these circles have often been dug out, (re)erected, and (re)arranged in circles from and in soft ground. In some well-documented cases, such as at Avebury and Stonehenge, the stones have been variously subjected to destruction, repositioning, conservation, excavation, and many reinterpretations. An extreme example is at Fir Clump Circle in Wiltshire, the stones of which were broken up and flattened in the 1860s, then entirely removed in 1969 to facilitate a motorway (Burl, 2000, p. 311).

Despite these vagaries, a basic circle shape is not in dispute, even where the significance of the circle is unclear. In present-day imagination (and possibly in that of their makers), they represent a shift in architecture from megalithic chambered tombs to open-air rings symbolizing "a change from darkness to light, from the dead to the living, from the grave to the sky" (Burl, 2000, p. 38). Many early UFO contactees were similarly motivated by signs, shapes and by the symbolism of a mystical illumination that exceeds grounded limitations and overcomes and outlasts the grave (Festinger, Riecken, and Schachter, 1956, pp. 33–57). George Adamski, for example, was the leader of a mystical "Royal Order of Tibet" in California prior to his UFO contact, and supplemented his UFO activities with psychic channeling (Lago, 2015, unpaginated); in the UK the Aetherius Society, founded by theosophist George King as a UFO-centred religion worshipping aliens as "Cosmic Masters," was expectant of a messiah coming in a flying saucer (Flaherty, 2011, p. 594) and held prayer meetings at Bronze Age stone circles and cairns to invoke the assistance of an inter-planetary parliament on Saturn.

The instability of the narratives of making, observation-production, and interpretation of both UFOs and Bronze Age circles (and the determination of so many of those attending to them to either assert or seek some final resolution of them) has led us to question the identities and qualities of their materials relative to the narratives around them. Were the materials (rocks, aliens, space travel technology) in fact independent agents? Or were they entangled in an "ideological apparatus" by which "ideas.... are material actions inserted into.... material practices governed by material rituals which are themselves defined by the ideological apparatus from which we derive the ideas of that subject" (Althusser, 1971, pp. 158–159)? Or was it with the very diagrammatical qualities of saucer and disc or circle and ring – with approaching/retreating from grave to sky, and saucer landing/rising, constituting a kind of time cone – that we could make an alliance?

In response to such uncertainties and affordances, we identified four distinct operations within the process of our devising and performance – substitution, intuition, amplification, and re-re-connection – and we will address each in turn at the end of this case study.

Bryant's story

Our *Scoriton Mysteries* project originated in a walk in 2015 with the ornithologist Tony Whitehead, to a village on the southern edge of Dartmoor. We were looking for anything recognizable of the sites of the two UFO encounters in 1965 experienced by Arthur Bryant as described by ufologist Eileen Buckle in *The Scoriton Mystery* (1967). Given that Bryant's encounters consisted of a momentary sighting and a contact encounter lasting a few minutes, Buckle's 300-page book is a remarkably thorough, if credulous, investigation.

What Bryant reports is that around 11 pm on the 7 June 1965, while at home in his isolated Hawson Farm cottage, some half a mile outside the cluster of houses at Scoriton, he heard a humming sound. On going outside he observed a blue-lit object cross the sky above his house; it made a sound "like the slamming of the door of a castle with a long corridor behind" (Buckle, 1967, p. 65) and then disappeared. Next morning, in the lane outside his cottage, Bryant found metal fragments that he thought were glowing. He subsequently searched the adjacent fields and found foliage withered and more metal fragments scattered (which he collected). He also found a glass phial containing silvery sand and a piece of paper with the words "*Adelphos Adelpho*" (subsequently translated as "brother to brother") written in classical Greek. All this was recorded in the Autumn 1965 Journal of BUFORA (British Unidentified Flying Object Research Association).

Subsequently Bryant re-contacted BUFORA to say that he had had an earlier and more profound encounter experience, on the 24th of April of 1965, during which he had conversed with aliens and that he was now ready to tell his story, which "will leave no doubt in the mind of the believer that a new era is opened up" (Buckle, 1967, p. 50). In this earlier incident, Bryant had taken a walk around 5.30 pm up the footpath from the village to the edge of the moor. On reaching the moor (or "Down" as Bryant calls it), he looked back towards Scoriton "and as I did so a large saucer-like object suddenly appeared from thin air over a field next to the lane." Rather than hovering in a stable manner, the saucer swings "like a pendulum" (p. 61), then settles above the field. A door slides open and three figures appear, dressed in what Bryant takes to be "diving gear" (p. 62), silvery in colour and making a sound like tinfoil. One of the figures beckons to Bryant with open arms, and Bryant climbs into the field with the visitors, who remove their helmets.

Bryant is astonished by the pointed foreheads of two of the figures, by their four-fingered hands and by their eyes that have a "vertical cat-like pupil" (p. 62). The third is of "normal appearance." This third individual introduces itself as "Yamski," explains that they are from Venus and that they are contacting Bryant "because you are of Romany origin – as I am" and that "he and his people were going to bring proof of a wonderful existence and life beyond our understanding," but warned of "forces from another planet" in the guise of poltergeists "which were taking people from this world for what he described as procreation purposes" (p. 63).

Bryant is taken aboard the flying saucer, which is largely empty but for a couch, across which is draped "a purple coloured robe with a red rose beautifully embroidered on the sleeve," and a small concave TV screen. Yamski explains that the saucer is powered by "Ideo-motor movement" (which is a term for meaningful muscular movement when the conscious will is disengaged, as in dowsing, Ouija board use or divination by pendulum). Yamski then promises to visit again and says, comfortingly, "Great things are done by those who are alone" (p. 64).

The three figures collectively thank Bryant, who jumps down from the flying saucer; its door closes leaving no sign of a door being there (as with the saucer in the 1951 movie *The Day The Earth Stood Still*); "the saucer went up to a height of ten to twelve feet and suddenly vanished" (p. 64). In the few minutes of this encounter, Bryant covers all four main themes of the ufo narrative: nuts and bolts saucer technology, benign space brothers, malevolent abduction, and inter-dimensionality. It is as much a diagram of the UFO myth as an account of an encounter.

While Bryant presents his story as an actual event – despite his wife's later description of it as being part of a science-fiction story he was writing – it can also be understood as part of "UFO theatre" (as other entities presenting as "aliens"), as a "blurred reality genre" in which the witness/experiencer "rather than reconciling scientific or religious notions of the world subverts them" (Dewen, 2006, p. 197), or as "crack experiences" (Rojcewicz, 1989, p. 12) that undo the experiencer and their ideas. The multiplicity of UFO accounts (and kinds of such accounts) and the vast range of world views and the psychological and philosophical reassessments they inspire have become a multiplicitous thing-in-themselves, resistant to being judged empirically by attention to their phenomena. Instead, they constitute an open and diverse making (and more often repeating) of narratives.

Based on his idea that "[T]here is no such thing as 'the UFO phenomenon' but there are lots of phenomena that cause UFOs," David Clarke proposes that the accounts of encounters "that form the core of the syndrome" should be regarded as "folklore" (2015, pp. 272–274). So, if we put aside binary questions of fictioning or veracity in the case of Bryant's story (was he telling the truth? Did he make it all up? Did he hallucinate due to a brain tumour?) and accept it as culturally significant folklore, how does it sit with other lores (mythic, historical and ecological) of Dartmoor? And how might that help us to perform them?

(Mis)Reading Dartmoor

On a recent visit to the site of Arthur Bryant's encounter with aliens, we passed the site and went through the metal gate there, onto a fenced part of the moor, along a "permitted path" (accessed at the discretion of the landowner). We knew from the map that there were burial cairns ahead that we were keen to visit. However, a fierce wind blew up and we grew nervous as the path was flanked by a line of huge beech trees, large branches from which had recently fallen across the path. Then we noticed, in a field beyond the trees, a large herd of red deer. Although they would have noticed us, only a few hundred yards away, they did not run.

The fencing, and our only partial right to use the path, was due to the deer being contained (there was nowhere for them to run to), probably bred for their meat or for hunting or both. The irony was not lost on us; for herds of wild red deer had been common on the moor until they were hunted to extinction, around the same time (late eighteenth century) as the last wolves there were reputably killed. In 1896, a paper given at the Devon Association noted the exceptional novelty of an appearance by red deer in Holne Chase Woods, five miles from Scoriton (Frith, 1896, p. 244).

Our assumption was that this centuries-old treatment of the deer was an example of a wild space eroded of its wildlife and only recently subjected to containment. On a subsequent visit, we went beyond the site of Bryant's encounter and up towards the summit of Huntingdon Warren on Buckfastleigh Moor, passing buildings named The Strole (a pinch point for controlling herds) and Bowden (a place for shooting with bow and arrow) and passed many geriatric oaks all of which are associated with deer hunting. Indeed, the name Buckfastleigh indicates the containing ("fast") of male deer (Gover, Mawer, and Stentson, 1931, p. 293) and the probable existence of a considerable Anglo-Saxon deer park, in use until a few hundred years ago; a pattern now being repeated.

By assuming a longstanding depredation to be recent, we had repeated a common misreading of the moor as a primal landscape that has only recently suffered but is essentially what it always was. In reality, Dartmoor's bleak-looking and windswept granite-based landscape, almost a thousand square kilometers in expanse, is not remotely like its ancient condition, which was that of a forest with isolated rocky peaks. Indeed, Dartmoor could be more reasonably interpreted as a repeat wasteland, subjected to extensive prehistoric deforestation, four thousand years of intermittent mining for ores, and the scraping of alluvial tin from stream beds; a setting for various failed agricultural projects over millennia. Recent depredations have included the removal of peat and an advancing moor grass monoculture in part due to extensive sheep farming.

Rather than an "untouched" landscape where a direct connection to prehistoric ancestors is readily available, Dartmoor is a terrain where meaning and materials are both unstable and enigmatic. Early on in the project, we chose to take the roughly circular shape of both the Bronze Age stone circles on the moor and of 1960s flying saucers as a diagrammatical invitation to explore their shared enigmatic quality. For the stones, despite their resilience (mostly granite) are every bit as enigmatic as UFO phenomena. The 90 stone rows and 15 stone circles on the moor recorded on the Prehistoric Dartmoor Walks database (2015, 2017) may look "ancient" in their present condition, but they have been subjected to centuries of destruction, extraction, and renovation. Circles have been raided and stones broken up by farmers seeking material to make gate posts, while at least one circle was seriously damaged during World War II by American troops who used it for target practice.

In late 2024, the discovery of two new circles was announced. It was claimed that these might have formed part of a "sacred arc" of circles "intended to ring the high ground in the centre of the moor" (Morris, unpaginated), implying that all or many of the circles might be parts of a much larger pattern. As yet, there is insufficient evidence to be anything like sure; individual circles are not consistently aligned either astronomically or to features in the terrestrial landscape. They are mostly sited close to the remains of prehistoric settlements, and while there are signs of large fires at circles that might indicate cremations, meetings, or communal festivals, evidence is inconclusive. The stones – like the UFOs – are open to speculation and affordant to "fictioning" as numinous artefacts and fugitive events.

Similarly, trying to fit Arthur Bryant's encounter experience within any one iteration of "the UFO phenomenon" is thankless; evidence is plentiful but everywhere compromised by its advocates' predilections. Our approach has been to accept no theory as comprehensive, but to reject no evidence or interpretation as meaningless; all being, at the very least, expressive of the subjective and social imaginations from which they come. We have stayed open to stories, objects, reports, and events, whatever their source, as possible agents in a meaningful performance.

Meanwhile, from the start of the process, we have considered the possible diagrammatics of the work. This means we have not been forced to choose some things and throw out others, but to place things differently within the emerging diagram, free to move them around without reference to any overarching theory. A pattern has emerged at least partly through the self-organization of its disparate parts (Johnson, 2001), freeing us up to attend to instances of alignment, such as a herd of red deer with old place names.

The hinge

In his book *Star.Ships* (2016), author Gordon White tells of the disruption of assumptions in academic archaeology and pre-history studies caused by the excavation of the complex at Göbekli Tepe in what is now Southern Turkey. A considerable (at least) ninth or tenth millennium BCE stone structure of large ornately engraved T-shaped columns arranged by hunter-gatherers in rough circles, Göbekli Tepe challenges (if not dispels) a prevailing view of hunter-gatherers as dull, hungry folk driven by necessity, and throws light on the sophistication of a Middle Stone Age community acting as philosophers-with-stone.

White rejects a "stages" theory of history – in common with James C. Scott in *Against the Grain* (2017) and David Graeber and David Wengrow in *The Dawn of Everything* (2021) – according to which Göbekli Tepe would have been built on the orders of priests sitting on surpluses of cereal crops, as an expression of a new technology or novel social system. Instead, White suggests "[W]e built Göbekli Tepe because it expressed something about ourselves and our place in the universe" (2016, p. 17), using "old-fashioned" Stone Age technology: "by hitting rocks with other rocks, then standing them up" (p. 22). In a sideswipe at pseudo-archaeologists (Graham Hancock, et al.) who have expressed or implied scepticism about the capacity of prehistoric and indigenous peoples to construct such impressive assemblages, insisting or implying that they had help from superior beings (from Atlantis or a distant galaxy), White writes "[N]o aliens required" (2016, p. 22).

However, what if in 1965 aliens *were* required? What if, unlike those hunter gatherers more than eleven thousand years ago, or the Bronze Age Dartmoor dwellers, in 1965, in the face of the threat of Cold War extermination, following two world wars, in the wake of colonialism and slavery, paranoid about radiation, what if Arthur Bryant could no longer express "something about ourselves and our place in the universe" without (at least welcoming, imagining, hallucinating or envisioning) outside "alien" help? And if not in 1965, could we now, in 2025, as Gaza is razed, with a war in Ukraine grinding its human corn, in the wake of a worldwide pandemic, subjected to ever more extreme climate-change weather, make a new performance in stone? Perhaps with the help of others, fictioned or otherwise; inviting the presence of "aliens" from distant galaxies, dimensions, species, or deep histories? But if so, how would we invite them?

The saucer-shaped hill and the pixie ring

While we might reach out to Yamski and their companions for our diagrammatical performance, or look to gather equivalents of the pieces of flying saucer cast into a lane and field in 1965, what of the makers of the Bronze Age circles, and their materials? Could (and should) we treat them as folkloric characters and properties too? Concerns about the appropriation of ancient cultures or those of indigenous peoples with identifiable millennia-old practices still in transmission, and the clumsiness of some previous revival attempts (anachronistic Druids at Stonehenge, for example) might encourage an understandably severe inhibition around such efforts. On the other hand, were we to rely on our subjective affects in the terrain around Scoriton? Or on a critical narrative that might connect us to a longer historical process on Dartmoor?

Unexpectedly, what stepped forward was the agency of grass. Our first visit to Scoriton, long pre-dating the *Scoriton Mysteries* project, had been part of an ambulatory research around South Devon to record its "mythogeography" (see Smith, Phil, 2010), eventually published under the pseudonym "Cecile Oak" in *Anywhere: A mythogeography of South Devon and how to walk it* (2017). In this book, Oak, a fictional narrator of actual journeys, describes our walk up the lane from the village (Oak, 2017, pp. 302–307), retracing Bryant's footsteps to the site of his encounter with aliens, where these days the path to the moor is metal-gated and fenced.

Finding the exact spot from which Bryant witnessed the appearance of a flying saucer, thanks to his vivid descriptions recorded in Buckle's book, Oak (and we) were able to observe the shaping of the terrain in which he was "contacted." Immediately behind the field, above which the "large saucer-like object suddenly appeared from thin air" (Buckle, 1967, p. 61), the ground rises up to a steadily sloping summit some 300 feet above sea level. Standing in Bryant's footsteps, we noticed how "the thick mass of dark treeline rises only so far up the hillside" opposite leaving a large "strip of bright green grassland under the sky: saucer shaped" (Oak, p. 305).

Given Bryant's description of the time of year and the hour of his walk, the direction of sunlight striking the top of the hill, a possible visual disturbance due to a brain tumour that would kill him shortly thereafter, was it likely that Bryant's flying saucer consisted mostly of the moor in the shape of the bright grassy summit of the sunlit hill talking to his imagination? Had Bryant translated his sublime apprehension of the moor into a ufological or sci-fi vocabulary of a hill speaking of alternative worlds – with a message about "a wonderful existence and life beyond our understanding" (Buckle, 1967, p. 63) – through the medium of his tumour?

While we had first come to Scoriton in pursuit of a more-than-human agent, the most likely and viable more-than-human agent turned out to be far more earthly and earthy than we had expected. Eight years later, returning to the same spot with Tony Whitehead, we noted other saucer shapes, including the shape of the trees on top of Denbury Hill hiding Iron Age earthworks that enclose two Bronze Age barrows. A year later, in early 2024, we returned to find another saucer shape performing itself. Passing through the metal gate to the moor, below yet another bright saucer-shaped hilltop, we spotted a dark circle in the grass, a "pixie ring"; an uneven growth of grass caused by the action of the mycelium of fungi absorbing nutrients in the ground, the fungi pulsing outwards from a central point (as it uses up the nutrients there) to create the ring. Could we make an alliance with this grass? There are precedents; in Iranian philosopher Reza Negarastani's *Cyclonopedia: Complicity With Anonymous Materials* (2008) – a theory fiction of the Middle East – the author makes oil the central character and driving force of his text. Cornish film director Mark Jenkin, in his film "Enys Men" (2022), makes lichen an independent force, attaching itself to a scar in the main character's belly.

The mycelium – the ultimate connective agent – of the pixie ring had invited itself into our performance, and us into its. We reciprocated by throwing our arms to the sky in the centre of the circle. By our "being there" and ready to perform, the place had begun to perform back to us with the same levels of ambiguity as ufology and prehistory. Returning a few weeks later we used white China clay powder from the nearby quarries to make a circle in the grass, and brought with us props (plastic aliens, flying saucer-shaped trash, red threads, and a small wooden statue of a dome-headed *lama* to represent the early mystic-contactees) to make a shrine to the ufo by the grass circle. Only small elements from these preparations – the pixie ring represented by a green ribbon, for example – would make it into the public performance, but all the time, material and detail involved served to create a "thick" performing presence.

Performing with a moor?

Now that more-than-human performers were stepping forward for inclusion, could we draw on recent folkloric tellings and doings on Dartmoor? There are certainly multiple examples of contemporary folkloric performances happening on and around Dartmoor, such as the annual wassailing and the raising of Old Crockern on Stall Moor in 2023 (see above), the storytelling of Sara Hurley and Lisa Schneidau in *Chasing Crockern* (2025), processions of Green Men, weddings in the circles, neo-Druidic solstice ceremonies to welcome the sun, and the narrative dances of Acid Morris and Beltane Border Morris. There are precedents: Cecil Torr describes how two women, in the early twentieth century, on their way across the moor to attend a church service, were approached by revelers from "some Carnival festivity, attired in costumes and masks" who joined hands and danced in a silent circle around the two women and then, without comment, processed off (p. 35). At the same time, there is a folkloric tradition of performance associated with Dartmoor's stone circles. The Nine Maidens circle is associated with a common story of being the remains of young women who danced "naked in the noontide sun" on the Sabbath (Torr, 1921, p. 36) and invoked the wrath of God. "Some stones in Devon turn at midday, including the Dartmoor stone circles" (Brown, 1964, p. 153); accounts describe stones dancing during heat hazes (Torr, 1921, p. 36) or moving like flocks of sheep in the case of the Grey Wethers circle.

Given a waning of commonsense incredulity about the venerable origins of folk practices, we felt encouraged to exercise caution before making essentially modern and European Enlightenment assumptions about linear time. Why should we accept the past-present-future of a "historical narrative of catastrophe" – particularly in relation to ecologies – when we might "read catastrophe backwards as anastrophe" in which a part of modernity is always "travelling backwards.... to encounter itself, in time, as another" (Ireland, unpaginated). In such a schema, crisis is a "convergence misinterpreted by humankind" (Plant and Land, 1994, unpaginated) that includes the possibility of multiplicitous alliances with more-than-human others and the reinvention of ancient performance from crisis-convergence materials as just the most recent of many reinventions. This implies "various feedback loops operative between different times.... [and] a particular kind of 'time diagram'.... that involves seeing time from.... a 'non-standard' point of view.... In which the future works back on the past" (O'Sullivan, 2024, p. 44).

In place of a theory – of UFOs or the origins of melancholy – Simon O'Sullivan suggests mobilizing fiction "in the war against time," using the tool of the "cut up" by which (following William Burroughs) reality can be seen as "a series of scripts.... which could be edited and rewritten" (2024, p.45). Here, in the context of *Scoriton Mysteries*, by applying the cut-up to ourselves, we sought to avoid a colonial-style plundering of culture. Thus our imagining of ancient performances and alien visitations (see figure 6.1), with

Figure 6.1 "Scoriton Mysteries." Crab & Bee at SOAK, The House, Plymouth (UK), 2025. Photo: Tamàs Kovàcs.

their explicit assumption of atavistic resonance, became a script we could "cut up," rearrange on a diagram and web with an ecological crisis-convergence.

We have no illusion that we are in some direct line of transmission from ancient performance. Rather, what we are doing – we might suspect – is what ancient performers might also have been doing: renewing their performance by editing and rearranging in the process of transmission; a process that "both keeps and transforms choreographies of meaning" (Taylor, 2003, p. 20).

In articulating his "'non-standard' point of view" of the operation of time, in terms of "remnants of the past that survive into the present" and "elements of the future that are operational in the present" (O'Sullivan, 2024, pp. 52, 51), Simon O'Sullivan, referencing Sun Ra's myth-science and the art of Black Quantum Futurism as relevant practices for shifting the direction of time, presents his diagram (see Figure 1.1, above) for the activation of a device "to follow different time circuits and temporal loops." In this diagram, the tip of a downwardly pointing time cone drags in the "past/virtual" from above into a sharp focus on a plane of matter and landscape while a loop of imagineering (a "[T]ime-travelling circuit") repeatedly accesses both the past and the point of contact with a present matter by looping up and down within the cone.

For the purposes of *Scoriton Mysteries* we inverted O'Sullivan's diagram, so that our imagineering drags the past up from the ground to a point of connection with an elevated plane of matter/landscape (mimicking the Bronze Age turn from grave to sky). At the same time, the shrinking or expanding circles around the cone (we used a traffic cone in *Scoriton Mysteries*) represent ripples of circular/saucerous performance revolving in their own present (of which one was our performance), echoing at different scales both past and future performances to which they are linked by imagination but separated by time and space.

Following Amy Ireland's model (taken from the shadowy, possibly fictitious scholar "Giraldus") of "two intersecting cones, one inverted and nested within the other" (Ireland, 2017, unpaginated), we proposed to nest our inverted model within Sullivan's to more accurately model our eventual devising process in *Scoriton Mysteries*. During that process, we identified four distinct operations within our process of devising and performance, made in response to the historical uncertainties and diagrammatical affordances mentioned above: *substitution*, *intuition*, *amplification*, and *re-re-connection*:

1. *Substitution* took place in the absence of a clear and certain tradition of performance. Without pretending to authenticity, we invented our own performance procedures in response to the ambience of the spaces we attended to, including sitting the audience in a circle on the stage area of The House studio, Plymouth.
2. *Intuition* included a repeated feeling out of the dynamics of the encounter sites, addressing the clues provided by the ambience, the feel of materials, and the impression of the shaping of the place, as well as intuiting what the performance should be when we were with the objects in a devising studio.
3. *Amplification* was present in performance through subjective anecdotes, striking "props" and materials (cardboard tentacle, black treacle), sensory invitations (to eat flying saucer sweets, hold pieces of trash against the head), and the ritual-like structure of the actions performed.
4. *Re-re-connection* was attempted to Arthur Bryant's original encounter experiences via the accounts in Eileen Buckle's book and other evidence from ufologists, the geography and ambience of the encounter sites and our own intuiting presences there. We sought to *re-re-connect* to the moor and its cultural heritage via the shape of the circles (stone and fungal), the intuiting of Bronze Age motives and local folklore, and by our repeated visits to the moor in general and the Bryant sites in particular.

Performance 4: *A séance for Stonehouse*

Context

In 2023, Crab & Bee made a performance that we called *A Séance for Stonehouse*. The idea arose from our *Plymouth Labyrinth* project (2018–2019) when we worked from and staged an exhibition/installation at a former pub in the Stonehouse area of Plymouth – The Royal Adelaide – which is now run as a gallery called RAAY. Little change to the building had taken place since it closed as a pub, with many of the bar features still in place and even some furniture remaining. The present-day owners, moving in, discovered many materials in the attic of the pub, including hundreds of postcards sent from all over the world to the pub by its patrons (on holiday abroad or from Merchant or Royal Navy sailors working or serving in distant ports).

During *Plymouth Labyrinth* we spent a lot of time in Stonehouse, often with local people engaged in rebuilding the area in both community and architectural terms. Stonehouse was devastated in the 1990s due to the collapse of its night-time economy of clubs and sex work in and around the notorious Union Street, partly the result of police actions against the drugs trade (including the confiscation of property and jailing of club owners). The depletion of Royal Navy numbers around the same time robbed the area of many of its customers. One of our poems generated during this project, in part, read:

> 'bleaching is hard to talk about
> when all that's left is a colourless ghost and
> an off the peg stone circle of granite gateposts
> redevelopment is
> a sterile wraith, a gated ghost dislocated from the dead
> who boned here....
> the playground was always chaotic once the toffs had been seen off
> now...
> ....a red silk dress hangs
> in an old morgue
> there are many things
> that need to stay interred,
> but others mourned until a red that's better, deeper, redder returns'.

Our intention for *A Séance For Stonehouse* was to perform some raising of a red ghost of the former energy of the place (acknowledging that this was not always a positive one) to aid its revival in the future.

The most immediate impulse to make a performance was an oblique one; Helen, following the *Plymouth Labyrinth* had wanted to place a permanent shrine in a public space in the area in recognition of its sex workers over many centuries (the last vestiges of its once considerable street sex industry having disappeared online in recent years). Although we located a suitable wooden cabinet for the shrine, we were never able to find a workable public place for it. Instead, we switched art forms to make a performance around the cabinet for the upstairs rooms in the Royal Adelaide. We chose to go back to the former pub after learning from local historian Richard Fisher during *Plymouth Labyrinth* how the Stonehouse pubs had doubled as brothels; pub landlords renting out their upper rooms to sex workers and taking payment from clients over the bar.

Properties

For the performance of *A Séance For Stonehouse* we used a number of individual objects: a pair of bright red high-heeled shoes, the cabinet (painted green on the exterior, the interior lit red), deer bones, old pennies and a large concrete ornamental farthing, a white glove, a photograph of Phil's half-uncle who drowned at Scapa Flow after embarking on the battleship The Royal Oak from Stonehouse in 1939, a pressure gauge and a number of power switches rescued from the pub's attic, a black wheeled stool left over from the pub, stones for the hardcore used to fill in Stonehouse Creek, a bottle of local Devon Red cider, red thread, and a flag-shaped charred piece of a chair found on a nearby Tamar river bank.

Devising the performance *in situ* in the pub rooms, we dispensed with most of the draft text we had prepared. We replaced it with the moving and placing of the above objects, repeatedly removing and returning them to the main room. The weave and significant placing of objects (eventually in the cabinet/shrine) became both a séance and an emerging diagram. This process was explicitly prefigured in an early draft of the performance text which begins with an explanation and invocation:

> Usually, when we've made these diagrammatical performances, we've used a performance area, like this, as a kind of map onto which we've placed objects to represent places, and tapes to represent the flow of stories or rivers or ideas or journeys, but Stonehouse doesn't like you trying to do that. At one level the place is far more alive and in its body than most places we've mapped; and at another level there is far more missing than is here; the Stone House is full of ghosts… A diagram of missing things is a séance. So, this is a séance.

The licking of a glove, repeated throughout the performance, we took from a local TV documentary made about sex work in Stonehouse in the 1960s, during which a stripper was shown licking her glove and shoe as part of her routine. The cripplingly high red shoes we bought from a local charity shop. At one level these objects and actions all had a direct hyper-realist connection to objects and history in the area; they either were the objects/actions or they acted as sympathetic substances connecting to similar substances elsewhere. Their secondary purpose was as markers holding down the elusive diagram of the séance; a third was their orbiting in relation to each as part of an orrery of attraction.

As part of their doing (by them and by us), they illuminated themselves across the web of connections and attractions around Stonehouse as parts of the hyper-objects that were the main concerns of our performance objectives.

The latter included a reconnecting to and remembering of the marshlands that were drained to build the area (still evident in truncated and culverted flows) and their connection to wider water systems, an economy (of sex work, drugs and dancing) that has been suppressed but not replaced, and the connection of the sex trade via the medieval hunting of deer (similar to London's Soho) to the extractive industries of clay and tungsten mining on the North-Eastern edge of the city. We repeatedly used the switches found in the pub's attic, clicking them on and off, no longer connected to anything, as a means to invoke the energies lost to the area, from the life energies of sex workers worn down by their trade to the flows of waters that once defined the area as an island independent of the city's control, once called "Women's Town" according to a local historian.

We sought to sympathetically reverse a process described by Timothy Morton in which massive objects or networks impinge directly on us – "a supermassive black hole.... as if it were sitting in the car next to me.... global warming burns the skin on the back of my neck" (Morton, 2013, p. 27). By using the small familiar objects and sensations of the site/non-site of our performance we connected to the hyper-objects of which the objects and sensations are or had been a vital part: global sex work, trans-national economies, depredated environments.

A note on preparatory actions

We have often brought lead, clays and clay powder, fabrics, treacle, stones, UFO-shaped gutter trash, machine parts, and deer bones into our performance spaces. When the places have not been accessible to us, we have recreated them in places of equivalent significance. So, for example, prior to using the black treacle in *Scoriton Mysteries* (2025), on the day after the death of David Lynch, we took a tin of the treacle to a grove of trees near to both our homes – our own "Glastonbury Grove" – and, after libating with bottles of red and white water taken separately from Glastonbury's ferrous Chalice Spring and limy White Spring, poured the black liquid into the ground there, partly in recognition of a debt to Lynch and partly "powering up" the material for our performance of *Scoriton Mysteries* a few days later. The implicit knowledge accreted within our bodies, accumulated through focused phenomenological exploration of the specific and connected places, and with the materials, was thus available to us as immanent potential for "vital acts of transfer" (Taylor, 2003, p. 2), through both our "archive" and our "repertoire" (p. 16), in a way that it would not be, had we simply stayed at home and read about the places. Or stayed in those places to make performances, in a conventional site-specific process.

If this all seems like an overly laborious approach to devising simple gestures – making diagrammatical performances unlikely to feature in an accumulation of cultural capital – then that is because we have not found rapid alternatives to finding the requisite thickness of experience and expression. In this sense, we make common cause with the monk-actors of the Grotowski tradition, not as regards their rigorous bodily training, but in the sense of the inefficient and labour-intensive daily nature of our practice of walking and "being there" with the subjects of our performances. Rather than our bodies, the grounds of our performances are the places, or to be more precise the *genii loci* and things of the places we spend time being with.

Understood through the prism of Morton's hyperobjects, the things (places and objects) we use in our performances are doubled, as in Henry Daniel's staging, not only as anchor points for the diagram, but more as hyperobjective energies that come rushing in from afar. As Timothy Morton warns, "'distance' is only a psychic and ideological construct designed to protect [us] from the nearness of things" (Morton, 2013, p. 27). In our performances, Crab & Bee's objective is to gently subject audiences and objects to the collapse of that "construct"; so in *A Séance For Stonehouse,* audiences were handed old copper pennies, rubbed smooth from a century in circulation, which they were encouraged to place in an empty beer mug as "something for the stripper," making them uncomfortably complicit in an economy of sex work. In *Scoriton Mysteries* (2025) the audience were asked to perform invocational gestures to call down aliens and add their energy to a "prayer box." In *Patterns, Diagrams and Place Stories*, the audience were given clay with which to make ancestral heads as they watched, connecting their personal lineages (familial or cultural) to the prehistoric laying down of clay deposits as the performance mapped material and cultural deposits across England to South Wales.

References

Althusser, L. (1971) *Lenin and philosophy and other essays*. Translated by Ben Brewster. New York & London: Monthly Review Press.

Anthony, D.W. (2007) *The horse, the wheel and language*. Princeton: Princeton University Press.

Bayer, O., Simmonds, A. and Welsh, K. (2017) 'Excavation Of An Early Bronze Age Barrow at Emmets Post, Shaugh Prior, Dartmoor'. *Proceedings of the Devon Archaeological Society*, 75, pp. 51–95.

Binding, C.J. and Wilson, L.J. (2010) 'Ritual protection marks in Wookey Hole and long hole', *Proceedings Of The University of Bristol Spaeleological Society*, 15(1). Available online at: https://www.ubss.org.uk/resources/proceedings/vol25/UBSS_Proc_25_1_47-73.pdf. Accessed on: 29.11.24.

Blackmore, S. (1999) *The meme machine*. Oxford: Oxford University Press.

British Geological Survey. (2024) 'Foundations of the Mendips'. Available online at: https://www2.bgs.ac.uk/mendips/home.htm. Accessed on: 28.11.24.

Brown, T. (1964) 'The folklore of Devon', *Folklore*, 75(3), pp. 145–160.

Buckle, E. (1967) *The Scoriton mystery*. London: Neville Spearman Ltd.

Burl, A. (2000) *The stone circles of Britain, Ireland and Brittany*. New Haven & London: Yale University Press.

Clarke, D. (2015) *How UFOs conquered the world*. London: Aurum Press.

Coulson, S. (2008) 'Response to "World's Oldest Ritual Site? The Python Cave at Tsodilo Hills, World Heritage Site, Botswana" by Lawrence H. Robbins, Alec C.

Campbell, George Brook and Michael L. Murphy'. *Nyame Akuma*, Number 26, December, 2008, pp. 2–3.
Coulson, S., Staurset, S. and Walker, N. (2011) 'Ritualized behavior in the Middle Stone Age: Evidence from Rhino Cave, Tsodilo Hills, Botswana', *PaleoAnthropology*, 2011, pp. 18–61.
Crab & Bee (Helen Billinghurst & Phil Smith). (2020) *The pattern*. Axminster: Triarchy Press.
Crab & Bee. (Helen Billinghurst & Phil Smith). (2025) *Crab & Bee's matter of Britain*. St John's Chapel: Peakrill Press.
David, B., Mullett, R. and Wright, N. et al. (2024) 'Archaeological evidence of an ethnographically documented Australian Aboriginal ritual dated to the last ice age'. *Nature Human Behaviour*, 8, pp. 1481–1492. Available online at: https://doi.org/10.1038/s41562-024-01912-w. Accessed on: 28.10.24.
Dewen, W.J. (2006) "A saucerful of secrets": An interdisciplinary analysis of UFO experiences', *The Journal of American Folklore*, 119(462), pp. 184–202.
Federici, S. (2004) *Caliban and the witch*. New York: Autonomedia.
Festinger, L., Riecken, H. and Schachter, S. (1956) *When prophesy fails*. New York: Harper Torchbooks.
Flaherty, R.P. (2011) 'UFOs, ETs and the millennial imagination'. In *The Oxford book of millenarianism*. Edited by Catherine Wessinger. Oxford: Oxford University Press, pp. 587–610.
Frith, F.H. (1896) 'Red deer in Buckland Woods', *Transactions of the Devon Association*, 28, pp. 244–246.
Gover, J.E.B., Mawer, A. and Stentson, F.M. (1931) 'The places names of Devon, parts I & II'. *English place name society*, Vol. 8. Cambridge: University of Cambridge Press.
Graeber, D. and Wengrow, D. (2022) *The dawn of everything: A new history of humanity*. London: Penguin Books.
Green, M.J. (1997) *Exploring the world of the druids*. London: Thames & Hudson.
Hawkes, C.J., Rogers, J.M. and Tratman, E.K. (1976) 'Romano-British cemetery in the fourth chamber of Wookey Hole, Somerset'. *Proceedings of the University of Bristol Spaeleological Society*, 15(1). Available online at: https://www.ubss.org.uk/resources/proceedings/vol15/UBSS_Proc_15_1_23-52.pdf. Accessed on: 29.11.24.
Ingold, T. (2010) 'Ways of mind-walking: reading, Writing, painting'. *Visual studies*, 25. London: Routledge. Available online at: http://www.tandfonline.com/doi/full/10.1080/14725861003606712. Accessed on: 11.2.225.
Ireland, A. (2017) *The Poemenemon: Form as Occult Technology*. Available online at: www.urbanomic.com/document/poememenon/. Accessed on: 21.11.24.
Johnson, S. (2001) *Emergence: The connected lives of ants, brains, cities, and software*. New York: Scribner.
Jung, C.G. (2002) *Flying saucers*. London & New York: Routledge Classics.
Kartsaki, E. (2016) *On repetition*. Bristol & Chicago: Intellect.
Lago, D. (2015) 'Messages from space'. *Michigan Quarterly Review*, 54(1). Available online at: https://quod.lib.umich.edu/cgi/t/text/text-idx?cc=mqr;c=mqr;c=mqrarchive;idno=act2080.0054.108;g=mqrg;rgn=main;view=text;xc=1. Accessed on 14.2.25.
Mason, J. Private email, received 1.2.25.
Morton, T. (2013) *Hyperobjects*. Minneapolis: University Of Minnesota Press.
Morton, T. (2022) *Spacecraft*. London: Bloomsbury Academic.
Negarestani, R. (2008) *Cyclonopedia*. Melbourne: re.press.
Newman, B. (1948) *The flying saucer*. London: Gollancz.
O'Sullivan, S. (2024) *From magic and myth-work to care and repair*. London: Goldsmiths Press.
Oak, C. (Phil Smith). (2017) *Anywhere: A mythogeography of South Devon and how to walk it*. Axminster: Triarchy Press.

Plant, S. and Land, N. (1994) 'Cyberpositive', in Matthew Fuller (ed.) *Unnatural: Techno-theory for a contaminated culture*. London: Underground, pp. 3–10.
Renfrew, C. (1990) 'Languages of art: The work of Richard Long', *The Cambridge Review*, 111, pp. 110–114.
Rojcewicz, P.M. (1989) 'The Folklore of the "Men in Black": A challenge to the prevailing paradigm', *ReVISION*, 2(4), pp. 5–19.
Sagan, C. (1995) *The demon-haunted world*. New York: Random House.
Scott, J.C. (2017) *Against the grain: A deep history of the earliest States*. New Haven & London: Yale University Press.
Sky, S. and Mcphie, J. (2015) 'A last wolf in England: The Wolf's Tale'. *Dark Mountain*, (7), pp. 223–230. Available online at: http://insight.cumbria.ac.uk/id/eprint/3336/. Accessed on: 23.3.25.
Smith, P. (2010) *Mythogeography*. Axminster: Triarchy Press.
Somerset Wildlife Trust. (n.d.) 'Ubley Warren'. Available online at: https://www.somersetwildlife.org/nature-reserves/ubley-warren. Accessed on: 25.11.24.
Southwest Heritage Trust: Somerset archive catalogue. (n.d.) *Records of St Cuthbert's Paper Mill and Wookey Hole Paper Mill, Wells, 1863 – 2002*. Available online at: https://somerset-cat.swheritage.org.uk/records/A/COT. Accessed on: 26.11.24.
Spector, N. (2004) 'Only the perverse fantasy can still save us', in De Grazia, M. (eds.) *The cremaster cycle. Matthew Barney*. New York: Guggenheim Museum, pp. 2–91.
Stenner, R.D. (1978) 'The concentration of copper, lead and zinc in sediments in Wookey Hole Cave, Somerset'. *Proceedings of the University of Bristol Spaeleological Society*, 15(1). Available online at: https://www.ubss.org.uk/resources/proceedings/vol15/UBSS_Proc_15_1_49-52.pdf. Accessed on: 25.11.24.
St Ledger-Gordon, R. (1972) *The witchcraft and folklore of Dartmoor*. New York: Bell Publishing.
Sullivan, S. and Low, C. (2014) 'Shades of the rainbow serpent? A KhoeSan animal between myth and landscape in Southern Africa – Ethnographic contextualisations of rock art representations', *Arts*, 3(2), pp. 215–244.
Taylor, D. (2003) *The archive and the Repertoire: Performing cultural memory in the Americas*. Durham, NC: Duke University Press.
Torr, C. (1921) *Small talk at Wreyland, second series*. Cambridge: The University Press.
UK Southwest. (n.d.) 'Ubley Warren Nature Reserve, Somerset'. Available online at: https://www.uksouthwest.net/somerset/ubley-warren-nature-reserve/. Accessed on: 25.11.24.
Wadell, J. (2018) *Myth and materiality*. Oxford: Oxbow Books.
White, G. (2016) *Star.Ships: A prehistory of the spirits*. London: Scarlet Imprint.
Williams, E.C. (2011) *Combined and uneven apocalypse*. Winchester & Washington: Zero Books.
Wookey Hole Caves (2024) 'Historic Stories of Wookey Hole'. Available online at: https://www.youtube.com/watch?v=iWgKnRRrxuI. Accessed on: 27.11.24.

7 Mini-handbook of techniques

Given that the main theme of this book is an argument for handing over agency to more-than-human others, simply drawing lessons from existing traditions of human performance – including those of eco-theatre and site-specific performance – is unlikely to be sufficient. To diagram more-than-human relations for performance some awareness of their likely instability and flux may be important. So, if you are intending to deploy any of the ideas from this book, you may need to do as much invention and reconsideration as application. Key to everything you do will be your attention to the more-than-human agencies. If you have come to this through a concern over conditions of accelerating climate crisis, then you need to find allies among those more-than-humans that are directly impacted. You will learn much more from them by intuition than we can help you with here. We will do our best, but everything we suggest should not be taken with a pinch of salt, but with an ocean's worth.

Getting started

Maybe the best way to get started on a diagrammatical performance is to not get started. Instead, go outside, go for a walk and then wait. Let the ideas come to you. Don't chase them; wander out with your senses open.

As you go, consider your "situation" as an art maker.

We are not only in a climate crisis, but a techno-informational cultural one too. Has any art, live or mechanical or digital, immersive or low or high, proved capable of addressing (let alone changing) the devastation of Gaza or the multinational slaughter in the towns, woods, and fields of Ukraine or the creeping horror of environmental devastation sweeping species out of their habitats?

In the late 1940s through to the late 1960s, members of the Lettristes, International Lettristes and Situationist International felt a similar lack in art and sought to undo the culture of genocide and self-harm that had produced two world wars. They attempted this by a radical disruption and hollowing out of cultural forms, cutting up and redeploying artistic and media artefacts for radical purposes. However, like the Dadaists before them, they struggled

DOI: 10.4324/9781003606673-8

with the paradox of making "anti-art" or anti-culture, any manifestation of which, no matter how angry, was quickly appropriated by commerce and media, and rebranded as a gingering up of the established culture. In response, the situationists developed the idea of an "art of living." Rather than dedicating one's life to making art, using art to make one's life.

So, get started by borrowing that idea.

In the gaps left over from keeping yourself alive – work, family responsibilities, caring, and so on – whether it is for a few minutes or a few weeks, allow yourself to be made by the "art" and "anti-art" of more-than-human others. Go outside. Or move around your flat (or wherever you live) as if it were outside. Stay still, let things emerge; the grain in the wood you never noticed before, the odours of moss, the patterns of bird flight, the tapping of raindrops, listen for the sounds of metals in mechanisms complaining, the symphonies in ventilation units, the wordless howls that wind speaks in trees and power lines. Sensitize yourself.

After a while look for shapes and patterns in the ground. Be suspicious. Imagine the strangest explanation and only surrender it for something more mundane if you find the proof for that. Everywhere is ancient land. Remnants of the dead lie under every footstep you take. No matter how "hypermodern," or apparently homogenized, there will remain some shaping made by a long-gone people, no matter how hard modernists or those utilitarians who never notice have tried to scrub it out. Rather than eradicate the pattern, their efforts only redistribute it. No matter how urban, there will be shapes and textures from hundreds of millions of years of formation; every limestone building is the tomb of billions of Devonian period sea creatures.

On Dartmoor every gatepost might be a menhir or from a stone circle, broken up by a farmer to serve a purpose; in the walls of medieval Christian churches there are Bronze Age menhirs visible, redeployed as lintels and doorsteps. In the fabric of ancient shrines are fossilized trees and meteorites. In the Fountain of Youth Park in St Augustine, Florida, look for the shadow of the Great House of Chief Seloy of the matrilineal Timucua people. Along Broadway in New York, trace the route of the Wickquasgeck people. In our city of Plymouth (UK), no one seems to notice the telltale shapes of a roadside Iron Age earthwork partially obscured by trees and a Boy Scouts hut next to the local branch of Morrisons supermarket.

Even where developers have pursued a scorched earth policy, they can only ever provisionally control the water courses. When storms come, follow the flows that return to seek out the old stream beds.

Paint or embroider maps of the layered patterns all around you, or draw them out on paper, or with chalk on the ground. Most importantly, develop an awareness of the diagrams in your head. Somewhere down the strata you will meet ancestors who would barely recognize you, if at all; yet we all have ancestors wherever we are, we share that ground with those who came *there* before us.

Some of those ancestors will be molten rocks, sands, lizards, nomads, fungi; things and people with which we share no detectable "bloodline," but we can (and should) still remember them as family, as parts of our stories. There is no reason not to feel a fondness for them, and their strangeness; no reason not to reach out to a living relative and *their* strangeness. Enjoying and reverencing the weird – the thing that is here and of here but does not quite fit or make sense – expands our identity and spreads our feelings.

If you see a pebble, or an animal bone in the woods, or a piece of trash in the gutter that speaks to you in some way, maybe take it home with you. Make a gathering of things on your mantlepiece, or on your bedside table; they will become characters in your dreams. Maybe in your performances. This is a salvagepunk operation. In the hypnagogic and hypnopompic in-betweens, of waking and falling asleep and sleeping and awakening, you can meet these characters, ask them questions, ask them what role, if any, they would like to play in your performances.

These are building blocks for an "art of living." But they are only notional ones. Anyone can construct their own quests, their own foci, fashion their own ways to pay attention, to stop and listen. The Parisian situationists were urbanists who thought the countryside was boring; in the UK, there is no real distinction between city and country as far as wildness is concerned; both city and country are equally under human management. But anyone who has ever been managed knows about the limitations, venal drives, and lack of competence hidden behind authority; all of which conspire to rough up the smooth exteriors of managed landscapes. In the UK countryside, you will be challenged by members of the local fox hunt, help sheep trapped in wire fences, weep at the flailing of the hedgerows in spring when the chicks are hatching in the nests, and goggle at the trashing of public rights of way. You will notice green algae colonizing the exteriors of brand-new buildings. Cracks in the façade of human control let in a different kind of power and vivacity; pay attention to that. It is your key ally in constructing an art of living you can draw into your own realms of dozing off and awakening, and finesse there, ready to perform.

Research

Get away from books for a while (but only for a while). Get away from theatres. Get away from screens. Diagrams are not only part of performance making. They are out in the world. Some are random, accidental, coincidental. Then there is the whole physics of forces – gravity, electromagnetism, the nuclear forces that hold tiny particles together – is their consistency diagrammatical? Can you perform with them? Is probability at the quantum level random? If this all sounds a little wobbly, then maybe that is how things are? If so, maybe the texture of your performances should reflect that?

What distinguishes a diagram from a pattern is how a diagram always serves some idea or system – even if it is only the layout of a wiring system – whereas a pattern need only have a relation between the parts of itself; though it may sometimes represent ideas or systems. So, you may want to attend to those diagrams that humans have left in the terrain – often based on existing physical landscape features – some of them thousands of years old. Or others far more recently made. They were constructed by ancestors, distant or recent, who intuited something about a place and wanted to pay special attention to it; you might wonder at their reasons. You might wonder at what their diagrams and patterns represent.

In some cases, such traces include some enhancement of a shape already found in a place; some naturally occurring feature of the terrain to which people attributed meaning. Perhaps they turned a shape into a diagram in reference to some abstraction or assemblage of beliefs. In some cases, this may have distant connections to the arrangement of stars in the night sky, sometimes to invisible things, occasionally to visible landmarks, stories, or personalities of that place. Or it may have been where to find food, or water, or where to rest during a long journey. Or combinations of all those things. Most of the time, however, we can only guess at these things, but it is worth seeking out the work of archaeologists on these places.

To get in touch with these places, it is always essential to "be there" with them, to go there and attend to them and what is active there; observe what grows, what lives, what changes. Put your own thoughts aside until the place starts speaking through them and attend to what is going on outside yourself. Sit still and maybe things will come forward to meet you. Stay silent for half an hour and listen to what speaks: birdsong, machines, winds playing aeolian wires, heartbeats, burbling water. Take a notebook; record your experiences. Lie down. Listen out for yelled words, barks, howls, whisperings, bird chatter, sirens, horns, overheard conversations, rumbles, warnings, whistles; save them for possible performance texts. Make poems and tell them to a friend. Paint, draw; you may be sketching the structure of your next performance, but you do not need to be doing that consciously.

Whether the patterns of these places are astronomical, aquatic, geological, architectural, fanciful, or mythical, they are all open to being performed. Similar to places associated with legend tripping (see the Legend Tripping section above), they share qualities with those places the philosopher Simondon calls "privileged points" (Simondon, 2011, p. 412); places of intense connection. See if they work for you. A diagrammatical researcher may want to approach these places (and their modern equivalents) as affordances – places of entanglements and distractions – from which you might draw on their shapes and place-stories (the formational stories of the place) to make your own "pattern" (Crab & Bee, 2020, pp. 189–197) and around which you might fashion your performances.

When you find such a privileged point – a place of convergence on your diagram – find a way to mark it. Place a pebble there, tie a scrap of cloth to a branch, draw an "X" in the dirt with your finger, pour a little water on the ground, or strike a match. Improvise.

You may recognize patterns most clearly in some ancient form, but there are always plenty of recent examples. At the smaller scale these might include medieval protective marks scratched into the walls of pillars and churches (or of caves like Wookey Hole, see above), while at a grander scale they might include modernist versions of these patterns, such as the urban layout of central Milton Keynes (UK) with its labyrinths, Midsummer Boulevard, medicine wheel, light pyramid and cathedral of trees. Or in abject spaces you may find such patterns on discarded tin cans and lost playing cards, in the ruins of sacred spaces, on old movie posters, in a series of miscreant fires set over buried rivers, or in the use of patterns and logos by corporations on their products.

It may be that in such places you encounter a dynamic pattern. Maybe in the work of turbating worms in the mud of an old quay or creek, in a gyre of plastic waste, in the over-stimulated algae around an outflow pipe, in the desperation of hungry herring gulls, in the moss taking over the roof of a building, or in surprising flocks of parakeets. By repeated observation, and some reading up, you may begin to diagrammatize these entities and be ready to perform with them, or with their representations.

Games

It all begins with games.

Once you have begun to connect with places and things and others, invent your own games. Draw on whatever games you know, from "catch" to baseball, tiddlywinks or leapfrog, chess or poker, geomancy or alchemy. Games are often the quickest way into diagrams. Can you use and transform the layout of playing fields and gameboards? Can you deploy and adapt the charts that tacticians and coaches use to train their players? Who are the players? Those entities like parakeets and moss and plastic waste? Why not? Can you move about, with them, like pieces on a chess set, or like the counters on a snakes and ladders board? Find ways to cheat; break the rules to make them better.

Maybe you can transform the lines and boundaries of your games into players, or "sides"? Consider the stabilities that playing gives to action, with its lines, its "in touch," "behind the goal line," "out of play." What happens when the lines begin to act for themselves? In considering these features of pitch and gameboard, theorist and performance maker Richard Schechner considered various terms: "'frame'.... it's too stiff, too impermeable, too 'on/off,' 'inside/outside'. 'Net' is better: a porous, flexible gatherer; a three-dimensional, dynamic, flow-through container" (Schechner, 1993, p. 41) – and so we are back with the carrier bag of Ursula Le Guin. In game form. Can you take the borders and demarcations of your games and turn them into a diagram that can sustain and gather together the different parts of your performing? Can you find flying saucer-shaped sieves, tattered nets, cans full of holes, or their equivalent to act with?

Can you invent your own shapes, fashion your own gaming pieces in wood, clay, or in imagination? Set yourselves tasks and quests? Go legend tripping (see the Legend Tripping section above)? Begin to "play" in the real world? Consider scale: find ways to play your game on a tabletop, then in the landscape and then across the cosmos.

In 2016, at Helen's *Crossing England* exhibition at the Ariel Gallery in Totnes (UK), the first time she had shown artworks generated from her walking between Plymouth and Cambridge, Helen drew in chalk a hop-scotch diagram outside the doors of the gallery. In part this was a reference to the chalk landscape around Cambridge, to her own childhood games, and to a connection between Cambridge and the legendary "leap" of Gogma-gog off Plymouth Hoe. In Helen's painting *Hopscotch/Giant's Leap* (2016), this "play diagram" is prominent. Children from the neighbouring school came to play on the chalked diagram outside the exhibition. Games thus became the transitional means for Crab & Bee to move from painting to performance.

When Crab & Bee first worked together on *Plymouth Labyrinth* (2018–2019), there were no plans for performance. Instead, the performance elements of that project grew out of spontaneous games; chalking and making labyrinths with red wool; the game of "swinging the bones" (very like a skipping game) became part of the "cabinet meeting" element.

Andrej Mirčev asserts in *The Poetics Of Performance Diagrams* that "the concept of the performance diagram creates a heuristic condition to rediscover and affirm precisely the ludic and ritualistic dimensions of performance making" (2024, p.71). The "good enough" but not fully optimal nature of heuristics (gathering knowledge by trial and error experimentation) is important here. For we are not deploying diagrams in a rigorously geometrical manner, but more in the way of the chalked (and re-chalked) approximations of a children's hopscotch game. That Andrej Mirčev includes rituals is also significant, partly because of the connection of games to rituals. For a while, Crab & Bee were careful to avoid mention or suggestion of any "ritual" aspect to our performances. We suspected there might be some resistance to the idea. Not that audiences might fear a supernatural presence invoked, but rather a feeling that rituals were for a particular milieu into which our spectators might not want to enter. We have since sensed a fading of this inhibition, but we still err on the side of the game and away from the ritualistic (partly because the latter implies something repetitive, and partly something elitist), but there is something there.

Rituals

One reason for our wariness of ritual is its potential to fix things, both in space and in meaning. However, these fixities can accompany flexibility.

Richard Schechner in his *The Future of Ritual* (1993) described (with diagrams) such an ambiguity in the spaces of the *Waehma* (Lenten cycle) of the Yaqui People of Pascua Pueblo (Arizona, USA) in the 1980s. Suppressed between 1886 and 1906 as a result of the genocidal diaspora of the Yaqui People, the *Waehma* was revived from memory as a constituent part of Yaqui identity. According to Schechner, "written into the theatrical-religious geography of the performance is a model of Yaqui history and experience as modern Yaquis have restored it," the performances display "great consistency" (1993, p. 97), serving a conservational purpose.

While nineteenth-century Jesuit missionaries had imposed churches and Catholicism on the Yaquis, the Yaquis interiorized their wilderness and magical pre-Catholic spaces – *huya aniya*, *yo aniya,* and *seya wailo* – and integrated them into the space of a *Waehma* Lenten ritual. Schechner detects seven definable spaces for this ritual, which lasts several days during "Holy Week." At times, movement in the spaces of the performance is rigorously controlled – "[T]he Loria Vo'o is strictly patrolled so that unauthorised persons – non-Yaqui and Yaqui alike – do not enter" (p. 104). The ritual cycle at times deploys the immediate layout of the town, while beyond the town a fifth space contains the "Anglo and Mexican world," which is spread throughout a sixth zone of desert, bushes, and mountains, which is "transformable into a *seventh*, the *huya aniya* (which space includes the *yo aniya* and *seya wailo*)" (p. 109). So, while ritual spaces are at times policed and controlled, preserving meaning, and restricting some spaces to certain persons, the ritual spacing can also be an arrangement of connections, a reaching beyond identity to an invisible zone or a far distance.

Thickness

There is something about repeated attention to certain things – sometimes apparently small or insignificant things – which somehow accumulates a "thickness" to them; an aura of portentousness. Occasionally, this is something – as with a found object – that we come to very late on in the process. This can be something, for example, like a part of an engine, that has been mined, forged, designed, cast, assembled, worked, broken, disassembled, kept, discarded; something which has already (by its dropping through and out of functionality) assumed a being-for-itself. In which case it is mostly a matter of keeping and curating the object sensitively.

Sometimes, the thickness is something we may have given an object from ourselves, or that has been given by those close to us. In *A Séance For Stonehouse* we used a photograph of Phil's half-uncle who died in 1939; we also used a pair of toy crows (with squeakers) that had belonged to Helen's son when a child. The worn pennies used in the performance had been in Phil's possession since being given them by elderly family members as curiosities when he was a child (they mostly date from the early years of the twentieth century). In *Leadworking* we used a gold ring that Helen had inherited from her father. Possession (in the theatre "props" is short for "properties") can have a thickening effect on what is "possessed" by some ambience, personality, or emotional attachment.

Sometimes this kind of attention may be something you have crafted yourself. In our case, we keep a number of sheep vertebrae threaded onto red and white woollen strings, which we can swing in such a way that it creates a "candy cane" pattern (as described above). This has a certain "life and death" ("boning" and bone) resonance for us. Its thickness is partly a result of making this simple construct entirely spontaneously, but it has accrued a thicker presence as we have "swung the bones" at poetry readings, in performances, and alone, for example on a (up to that point fruitless) search for the village of Maypole on the Isles of Scilly. The bones sit in a plastic lunchbox, sometimes they are nestled in lines from cut-up poems; we know that if the time is right we can take them from the box and they will hold and exude the thickness of their history (with us and long before). They are more than bones on strings, but it is enough that they are.

The principle of thickness means that we are less interested in the look or the representational qualities of a thing. A rusty railing will do for a weapon, a concrete garden ornament formed as an enormous farthing (rescued from a roadside, where it had been discarded) serves for a vanished economy. It is thickness, the quality both of the thing itself and of its relationship to us and its history with us, that imbues it with performativity and overrides its representational meaning. As with the objects in Tadeusz Kantor's theatre (Cricot 2, 1955–1991, based in Kraków, Poland), which serve as models here, they do "not appear to be supporting the action or the virtuosity of the performers," but rather they are "virtuosic themselves" (Allen, 2016, p. 10). We strap things to ourselves (a rusty key from the tungsten mine to Helen's forehead in *Leadworking*, Phil holding a flying saucer-shaped piece of metal trash to his head in *Scoriton Mysteries*). We also strap ourselves to things so that our movements are changed by the object, or so the object – including a costume – flaps or flexes autonomously; so during a single performance, a black academic gown references Hypatia's scholarship, the witchy darkness of Wookey Hole and the tarry feathers of Bran the Crow. There is an equivalence in using ourselves as agentive things; so we use a dustbin as a stencil to make (with the ash of our burned poems) a sun shape, but we also use our own bodies as stencils to make (also with ash) the outline of the body of the fallen giant Gogmagog. The thickness of these materials – ash from our burned poems, a rusted and holed dustbin lid found on threatened common land – is essential for their agency.

This thickness is dynamic. It is not itself a property, but a being. It is not an intentional communication with the audience, who will not know of it, and are unlikely to consciously recognize the objects' histories, although it is possible that their bodies may respond to the affective qualities of such objects. Thickness and affect are interrelated. We sometimes recognize things as actants, as an agency – equivalent to that of any human – that "modifies another entity in a trail" (Latour, 2004, p. 237); and this we do hope to communicate to our audiences. So, in *Scoriton Mysteries* we marked the ends and beginnings of some sequences by facing each other across the circle and simultaneously opening and slamming together a copy of Eileen Buckle's 1967 book *The Scoriton Mystery* and two pieces of wood hinged together like a book. There were multiple levels to this action – we were "closing the book" on the mystery, the loud slam/clack of the objects asserted the materiality of the mystery, and the woodenness – wood as a source of paper – hinted at the fabricated nature of the narrative we were exploring ("book" and "beech" having the same origin, we were also hinting at ambiguity, and at the shared origins of the disparate things in the Scoriton narrative).

So, choose carefully the things you work with. Learn to recognize when a terrain is giving up something for you. And when to leave it alone. Never take everything; always leave something for the place. Take things with you, ready to exchange one thing for another. Be ready to use your own possessions and to recognize which of them have the requisite depth of association and history. Be prepared to return things to the places you found them.

Icons

These are not mimetic representations of anything, but a kind of idol you can make; an object which can be – briefly – the material manifestation of the ideas to which your performance diagram refers. In *Leadworking* one of the ideas on the performance diagram was a syncretic relation of Christian saints with something older and more creaturely from the moor. We used a painting of Helen's, which combines an assembled "hunter/hunting ground" presence; melding a wolf head with the body of Saint Hubert, his two fingers raised in the gesture of "the speaker." Thus Hubert becomes a voice for the voiceless presence of wolves once present on the moor. Traditionally, Hubert is patron saint of werewolves (and their cure), hunters, woodsmen, and animal rights activists. Similarly, the clay figures of ancestors made by audiences *are* the idea of ancestors; they insist on a certain respect. Find ways for your audiences to make their own icons during your performances.

Afters

Just as a thread of ribbon ran from the stage of the Royal Conservatoire of Scotland for our performance of *Nets, Webs & Carrier Bag*, so you can expand the performance diagram by providing audiences with ways to connect to and explore their own lines, nodes and basins of attraction across the landscapes you have been invoking. To this end you can make and distribute small pamphlets or zines with a requisite map and any other imagery or texts you think will help extend the performance for audiences to a closer engagement with more-than-human others. Or distribute some other kind of guide, toolkit, or set of practical tactics to enhance an audience's encounter with their more-than-human allies. We like to send our audiences home with these, but we also distribute them more widely: placing them in telephone boxes, churches, libraries, community hubs, bus shelters, and niches in walls.

Probe-heads

In the 1979 Russian movie *Stalker*, directed by Andrei Tarkovsky, two strangers – The Writer and The Professor – are guided by the title character around a zone enchanted by aliens. The Stalker ties metal nuts to lengths of bandage and tosses them ahead of the party as they move through the zone. The effect is to make the visitors pause, causing a stutter in their journey, before moving ahead to the enchantment beyond the ruined tanks and abandoned buildings. Those bandaged metal nuts are versions of what Deleuze and Guattari identified as "probe-heads (….guidance devices) that dismantle the strata in their wake" (1988, p. 210). They delay the arrow-like lines of desire of the characters of *Stalker* (as they head towards a room that grants whatever a person desires most), which helps them realize that they don't yet know what their desires are. After the probe-heads are thrown the journey breaks up and the things of the zone impose themselves on the movie; the camera is no longer telling a narrative and instead moves over saturated objects and watery surfaces.

Probe heads do not only work on empowering the more-than-human others of Soviet film locations: all you need are some metal nuts and lengths of bandage. Or you could just use a small pebble. When it is time to question your performance, toss the probe-head ahead of you – "throw a stone over the wall" – and see what comes back at you.

A probe-head is only one of any number of similar fabrications you might make. They materialize an idea, in the same way that a diagram embodies a system or a philosophy. You can use them in your devising, or in the performances themselves. They are not substitutes for your materials, but they are there to enable your materials by putting a stutter into anthropocentric assumptions and returning focus to the crisis at hand.

Diagrammatical dramaturgy

Drawing from the distancing and cooling *verfremdungseffekt* of Brechtian Epic theatre, in which a reality is presented for inspection not represented for consumption, we can maybe create a "*Zustand*…. literally a standing-towards," a pause in the performance "picture," like those tableaux-like images from Symbolist theatre. In such moments, there is what Sven Steinhauser calls a "gesturing towards some other place and time" (2016, p. 25). Riffing off Walter Benjamin's account of Epic theatre, Steinhauser argues that by cutting away from a linear continuity, using interruptions and disruptions in what is expected (which is how Benjamin interpreted Bertolt Brecht's *verfremdungseffekt*) we can dismantle the conformity of "the more or less seamless passing of a plot towards its final conclusion." Steinhauser sees a parallel for this disruption in the clumsy changes from slide image to slide image in a Berlin magic lantern show called the *Kaiserpanorama* that Benjamin recalled from his childhood (2016, p.25).

In diagrammatical dramaturgy it is thus a virtue to expose such seams or the clumsiness between sections, to move jerkily sideways (horizontally) from one action to another, jump section to section, and allow gaps to open up between them – both in time and rational meaning – into which an audience are challenged to think for themselves. Rather than the conventional landscape of drama, with its sequence of increasingly tense climaxes and temporary relaxations, shaped like a seamless mountain range of peaks and troughs, a diagrammatical dramaturgy, once it has established its extra-ordinariness, is flattened, sustaining its intensity without seeking to bring it to any climax, final or provisional. (Any interruption of the intensity is likely to be a gap rather than a climax.) This has something in common with theory-fictions, there being "something radically democratic in their 'dropping down' or flattening procedure" (O'Sullivan, 2024, p. 9); the radical nature of such democracy being the access and power it accords more-than-human others.

To add to the meaning made by bumpy transitions in the dramatical structure, there can also be a kind of stutter in the gestures and language – "making language itself stammer.... placing all linguistic, and non-linguistic, elements in variation" (Deleuze and Guattari, 1988, p. 109) – or a shudder in the structure. At the same time, to avoid any all-too-easy consuming of immersive experience, a dramaturgy should render the audience experience questionable to its audience, not necessarily by violent disruption, but by introducing an uneasiness about the value and efficacy of everything. Challenging the audience to make a meaning from available ambiguities.

In some ways this should make dramaturgical construction much easier, without the need for any overarching structure for sustaining thematic and stylistic continuity and delivering believable but transformatory character arcs. In other ways, it is harder; for in place of those conventional structures, a diagrammatical dramaturg needs to find some non-rational, sensual, and ambient pattern (that appears in no toolkit) by which to stitch one section to the next. It is thus to the more-than-human others that we turn in these gaps; it is by attention to their flows, hunting lines, or patterns of distribution that we can hope to devise a creaturely dramaturgy.

Part of the point of such a non-rational, sensual, and ambient aesthetics is the unique fabrication of a "world" of disparate qualities and characters, with the aim of getting beyond novelty and looping back to "dull matter" (Bennett, 2010, p. vii). In *A Séance For Stonehouse* (2023) we switched from the "priestly" adorning of a shrine with vibrant objects to then imitating late twentieth century TV comedians The Two Ronnies (donning their trademark thick rimmed spectacles), then hopping abruptly between autobiographical anecdote and poetry to some dialogue taken from a 1960s TV documentary about local sex work. As we disorientated the performance, the old house began to perform with us; the audience told us later how as we briefly left the room, with no hint of a draught, dust gathered up and rolled slowly from one side of the room to the other. Before another performance of *Séance*, an audience member brought, as a gift, a wind-shaped stone, which we added to our "props shrine." The trick here is in shaping affordant gaps in such a way that things step in. Eschewing the usual peaks and valleys of conventional performance narrative, you must instead sustain energy, interest, and intrigue across planes or fields of action. You are not seeking to build tension to a climax (or a series of climaxes), but to quickly establish a heightened way of being – this can be done by visibly preparing the space, in a liminal time before anyone has acknowledged that a performance has begun – and then sustaining it until it is time to undo the performance and come down to a more equitable and sociable state.

Devising with more-than-humans others: The objects in the room

While you may wish to plan some kind of structure, make some things, gather some costumes, turn a walk into a drawing or a painting, learn some gestures, or write some texts as preparation, it is only when you are in the room with the objects that the devising begins. Unlike most rehearsals or devising for theatre, where the objects are worked so they fit a dramatic action, in diagrammatical theatre the devising process is for discovering the diagram made by the assemblage that is already in the room.

So, lay things out. Carry things from place to place in the room or space; where do they feel best, where do they fit, where do they react? Move things around again. They may move around without your volition, pushed by draughts, expanded by sunlight, changing appearance as well as position; they may begin to react with other things, creating their own display and reflection with and in each other. Move things around once more; can you sense which objects want to be with each other? Where in the space do the things want to be? How do they seem to work with your ideas for actions?

Stand back. Leave things alone for a while. Just watch, listen. Is there a more-than-the-sum-of-the-parts now present? What does touching, holding, and carrying the objects tell you about how to move? Are there ways that you can let the objects direct you? After a while, begin to make decisions about where in a performance space these more-than-human agencies most happily move and rest.

Diagrams in devising

Drawing diagrams – best to make them in freehand, to encourage their creaturely aspect – can help with devising in two respects. Firstly, they can help you to develop ideas in both an abstract and material form; so you can create pictures of fields of reasoning, or of forces at work on your themes, or draw a network of points or nodes, and use arrows (vectors) or ripples (waves) to show the flows and connections between them. The second respect is in regard to the transition from ideas to a performance. Diagrams that begin in the first aspect can move into the second as physical objects. So, for *Scoriton Mysteries* a cone shape scribbled in a notebook (denoting a funneling of energy from a broad plane down to a single point) would eventually – via a story about a UFO cult creating a "cone of power" in a stone circle in 1965 – become a physical traffic cone. In fact, it was a redeployed conical traffic bollard that we rescued after someone had thrown it over a park wall. There is a sort of bathetic humour here – of using everyday or abject objects to denote "high" conceptual systems – but it is also a provocation for an audience to see in everyday life that "ordinary" objects, usually ignored, carry concepts.

Diagrams are also a means by which to respond to what you are discovering in devising with the objects. A diagram can summarize what you have arrived at together and may – because the diagram expresses the dynamics of what you have – suggest ways to develop that. A diagram may expose a point of blockage you need to free up, or the emergence of something you want to encourage and highlight.

The "Cabinet meeting" model

This was invented by Helen for the opening of her *English Diagrams* exhibition at the Royal William Yard, Plymouth, in 2018. "The word cabinet refers to a box or cupboard but can also mean a private room for a council or a meeting place for advisors. A place to assemble or draw together a range of 'voices' or elements" (Billinghurst, 2018, *Cabinet Meeting* event flyer, self-published).

Guests and visitors stood amongst and around elements from the exhibition arranged on the gallery walls and floor in the orientation of Helen's walking from Plymouth to Cambridge; the artworks in the exhibition had been made in response to this journey. At the centre was a "medicine cabinet," a wooden box painted white with a bright red cross on its lid. Guests read aloud selected sections from a text about the journey provided by Helen. During the reading, there was some moving of the objects and the medicine cabinet was opened, revealing chalk and the bones of a white deer, found in (formerly Knights Templar) woodland near Cambridge. This ritual action sought to "redefine England"; from a "bounded, fortified" state, to an "open system, a crossing, or an enchanted space to be traversed and experienced" (Billinghurst, 2018, *Cabinet Meeting* event flyer, self-published).

We made a similar "cabinet meeting" for the opening of our *Plymouth Labyrinth* exhibition in 2019. During this reading one of the participants struggled with the text and their contribution was more of a sound poem, while another assumed a sinister character to deliver a curse upon a mining company (it worked, the company went bankrupt; and now the company that bought it up is going the same way). The text became more object-like. The content and literal meaning of a text are not necessarily the keys to such a "meeting." It is the speaking together within a circle that is key.

For our performance *Around The Hills* (2021) – in a garden on the cliffs above the Dorset (UK) coast – we expanded and dispersed the central cabinet to a series of similarly meaningful "stations" set out around a circle, with the audience sat or stood around half of the circle. In each case a "station" represented a place – a city, a hospital, an ancient settlement in Wales, the Tower of London, Plymouth Hoe – that was some part of a story web that we connected around and across the circle during the performance. Rather than have the audience read texts, we gave them clay to mould; first to shape bodies and then, separately, to shape heads. Then everyone paired up a body with a head. The combinations were arranged together on a garden bench.

Audiences

The position and "staging" of the audience is as important as anything else performed. Seating an audience in rows, as if expectant of a public lecture, will define their experience as much as any visceral or subtle effects from the performers. Free of commercial necessities, you can afford to have your audience small in numbers and close to you. Having the audience close up means that you can speak with more of an everyday voice, rather than "projecting" as an actor playing to a large auditorium must.

The most efficient way of having everyone close is a circle (see "Cabinet Meeting" above). If you choose this arrangement, however, you should remember that when you are inside the circle, you always have your back to someone, so you will want to keep moving around so you do not "blank" sections of the audience for any extended period. (The same applies, but to a lesser extent if you perform in a semi-circle or with an audience on three sides.) In *Scoriton Mysteries*, we kept a small space clear behind the four "stations" at the compass points around the circle, so at times one of us could join the circle by standing behind a "station," thus avoiding blocking the view of any of the audience.

More important than such technical requirements, however, is that the relation of performance to audience should be part of a diagram; circle, half-circle, rows, levels, or whatever you choose, should be parts of your overall diagram for the performance.

The role of an audience in diagrammatical aesthetics is not to interpret, nor to immerse, but to join in, to ally with the things of the performance, to marry the wolves, to play-pray into the "prayer-battery" (in our case a self-evidently empty plastic box or treacle tin), to model with clay, to gesture welcomingly to the alien others, to undo themselves as spectators. To turn around Baz Kershaw's claim about contemporary theatre that "any questioning of the… role of audience/spectator…. Tends to stop… at the exit or foyer door" (2007, p. 305). In diagrammatical performances we send the audience out with maps to visit original sites, with "probe-heads" to toss, or carrying ancestral heads with an instruction to find a place to enshrine them.

Spaces

The key to the space is the diagram. What the "three-dimensional metaphor" of a "non-site" affords us is a space "to-the-side" of the precise particularities of site-specificity: adjacent to the *genii loci*, history and fixtures of a place. A side-step which allows in the diagram. This "three-dimensional metaphor" is a model for performance-making (Smith, Phil, 2025, pp. 62–67) that we have derived from Robert Smithson's idea of a "site/non-site" interaction in which materials and information are moved from one place (of encounter) to another (of performance/presentation), the latter operating as a "three-dimensional metaphor" for the former.

The process here requires the uncomfortable extraction and displacement of materials from an original site (though one can usually return them), along with an "expressive" fabrication (including performance) of a space of "physical metaphorical material devoid of natural meanings and realistic assumptions," (*i.e.* the non-site) to which audiences can pay "a fictitious trip" (Smithson, 1996, p. 364). So the artificiality of the non-site can be used to its advantage as a form of *verfremdungseffekt*, deterring a tendency in site-specific performance to assume the authenticity of a site. Instead, by attention to the deep-time "pattern" of the original places (Crab & Bee, 2020, pp. 195–197) we can bring a flattened sense according to which any object in that place – terrain, person, monument, horse, tree, rock – is no more a subject than any other, but each valued according to its effects on others by the criteria of a web (DeLanda, 2013, p. 47). It is this relation that we attempt to transfer to and perform in the non-site. This "relation" becomes the diagram of the performance, on which points and trajectories are distinguishable by their mass, speed, and distribution across a performance-plane of consistency, sustaining differences across its structure (Deleuze and Guattari, 1988, pp. 78–79) rather than subjecting them to an internally consistent narrative or thematic plan.

"Non-sites" are not "empty spaces" in the sense that Peter Brook intended, meaning an affordance for performance, though they may be "non-places" in Marc Augé's sense of "transit points and temporary abodes" (1995, p. 78). They are spaces that are set aside in some way; they may have some history of being emptied and filled and emptied again. Given this adaptation to comings and goings, they are ideal for forming a diagram. We have found ideal spaces in a former pub, a theatre studio, a former paint factory, a cliff-top garden, a warehouse…. The "former" seems to be important here; anywhere that identity and specificity are rendered provisional.

In certain circumstances there may be no need to shift from the place of research to a "non-site" of performance. The original place may be quite diagrammatical (and accessible) enough. In this case, there is an overlap of site-specific and diagrammatical performance. In Phil's performance for the Glandwr Studio of *Starting Point* (2023) in Chapelizod, Dublin, the flow of the River Liffey (its splitting up and recombining in the village) and its interweaving with streets, paths and bridges already gave the place a diagrammatical quality somewhat akin to the meadow meander of Baz Kershaw (see below). There were also small nodes or "non-sites" (a cromlech in the park, a pub backyard) that were folded into the pattern. These Phil could then supplement with diagrams in chalk and clay on the pavements, inspired by the alchemical diagrams of Maria Hebraea of Alexandria and the personification of the river as Anna Livia Plurabelle; both referenced in James Joyce's *Finnegans Wake* (1939) which is set in Chapelizod.

Diagrammatical performances, given their relation to abstraction, and not being fixed to a single place, are able to shift around; they are nomadic. For example, Helen took the *Life & Death Walk* that Crab & Bee had made for St Mary on the Isles of Scilly in 2019 to Leo Camp in Norfolk (UK), a few months later; using the layout of hedges, fields, streams and bridges in place of the tracks, field gates and rock formations on St Mary. These performances are not specific to any single "site," although they may be intimately drawn from one. It is this nomadic quality, and their comfort in a "non-site," that perhaps explains how they can operate as both "folk" ceremonies with general communal participation, whether mediated or not by celebrants/performers, and as postmodernist performances in marginal spaces. In both cases they resist commercial priorities, they fail to meet the scaling up that "requires that project elements be oblivious to the indeterminacies of encounter" (Tsing, 2015, p. 38) and the reproductions required to generate profits.

Choosing a performance space

Choose your space carefully. Accept all ethical invitations. The best is when you can devise in the same space as you are to perform in. This does not oblige you to make a site-specific performance – you can still make a performance web ("non-site") to other (originating) places. But you should explore any performance space for its own qualities. What diagrammatical or scenic qualities does it have that you can work with? Talk to the owner or caretaker; glean what they know of the history and mechanics of the place.

When we worked at the Royal Adelaide Arms (RAAY), the former bedroom or living room that we used had a small room at one end, in which was a frosted glass window, at which we had puppet crows appear. Their portentous role, their comic toy-materials, mechanical squawks, and their blurred images in the window together made for a grotesque effect. In 2021, performing *Nets, Webs & Carrier Bag* at the Royal Conservatoire of Scotland (where there were any number of restrictions placed on what we could do in terms of materials and making a mess) we were able to run one of our threads off the stage and into a large storage area, thus, despite the prohibitions of this formal performance space, we could connect the performance to its wider web.

How can you get the diagram of the space to serve the diagram you want to perform?

In the case of our *Scoriton Mysteries* the answer was fairly simple. Because of everything about the subject matter, we wanted to use both the circular shape of the "flying saucer" and a cone (for the "cone of power" raised by the Aetherius Society). SOAK Live Art had moved their operation from an old paint factory to the modern studio theatre of The House in Plymouth, so we could bring all the audience to sit or stand in a circle on the large stage area, with the action and audience lit within a large single pool of light. This meant that we could do many things that an auditorium prohibits. For example, at the end of the performance we could use a thick green ribbon, for the "front row" of the audience to collectively hold, to form a green "pixie ring." This enabled the performance to shift its circle shape from a mechanical (or magical) saucer shape to a shaping in the fungi of the soil, in the grass, and in the hills.

In our *Jubilee Scrying* (2018) for artists collective tactileBosch in Cardiff (at an evening celebrating the 40th anniversary of Derek Jarman's movie *Jubilee*) we tied the audience together, at the wrists, with red ribbon (as red hunting lines of desire, and a nod to the image in the movie where a character is tied to a lamp post with barbed wire), before inviting them to scry (given the role of the magus John Dee in Jarman's film) in each other's eyes. Such repetitions with variations help to build the "thickness" of the action (see the Thickness section above and the Repetitions section below) to match that of the objects.

Nomadism

You will need, like any good nomad, to be able to defend your space, "non-site" or otherwise. You are always "on the road" and vulnerable. The diagram can help you to define your space of performance and protect it. At the same time, those who may think they also have a claim to the space may seek to take that from you; an organization hosting you may seek to hem you in with requirements and prohibitions. The more prestigious and "swanky" the venue, the more this is likely to happen; partly because there may be precious materials for them to protect (expensive wooden flooring, for example). This is the advantage of working in former industrial sites. Their ruinous or rough nature allows you to work more freely with materials that are likely to leave some trace. At the same time materials in that place may be breaking free and keen to work with you (don't ignore them). Working with the affective abjection of a space can help to sensitize, stimulate or "switch on" an audience in a way that is not possible in the becalming comfort of the black box space.

How to act 1: non-matrixed

Here "act" means something more like "behave" rather than "creating a character." Allan Kaprow and Michael Kirby both describe an approach to performing that the makers of "happenings" settled on in the 1960s. In Kaprow's case, it was something akin to a sportsperson who prepares themselves and their equipment, is well-organized and aware of spectators but not primarily concerned with entertaining them but attends most of all to effectively carrying out tasks and pursuing objectives within set rules and tactics (Kaprow, 2003, pp. 63–64). Michael Kirby characterized this mode as "non-matrixed," beyond the determining force of a "matrix" of texts or theatre customs. A "non-matrixed" mode thus allows a performer to generate and carry out actions that operate beyond or between conventions, unconstrained by traditions or common practices. This is helpful to a diagrammatical performance in that it aids us in resisting any habitual return to the anthropocentric assumptions of drama.

Kirby characterized a "non-matrixed" approach as being similar to how stage-hands might appear on stage in full view of an audience to move scenery and props during a break in the action. Or how wardrobe staff might enter the playing area to help a performer with their costume. Such a "non-performer" is aware that they are being watched, but they are not trying to convey any particular message or motivation or represent anything other than themselves and to complete their tasks as effectively as possible (1972/1995, pp. 40–41). In this way, storylines, objects, and materials can remain the focus of our performances, and we are their facilitators.

Like everything else in this toolkit, the advice in this section does not apply all the time. Preparing *Scoriton Mysteries*, following the guidance here, Phil made an error of judgement. In preparation, after pouring the sticky black fuel into the white flying saucer-shaped ceramic bowl, Phil advised Helen against raising it up and instead to carry it without fuss to a member of the audience. During the performance it became clear to us that we should have made much more of this moment, raising and processing the white bowl full of black fuel around the circle.

How to act 2: fungal gothic

While Crab & Bee have attempted a buttoned-down, non-matrixed stage-hand-like performance, Ali Matthews and Company's *Mushroom Language* (2025) took a very different approach (see *figure 7.1*); exercising and dismantling a range of extrovert performance styles, before bending round to an unexpected diagrammatical denouement. Their performance suggests that there may be any number of ways to perform with diagrams that we have yet to discover.

As *Mushroom Language* begins, the lights rise on a conventional performance studio space with a doubled version of the setting for Samuel Beckett's

Figure 7.1 *Mushroom Language: A Fungal Gothic*, Ali Matthews & Company, 2025. Photo: Georgiana Ghetiu.

Waiting For Godot, with two bare trees (rather than one) made of real branches and twigs, in a wasteland. The stage cloth, under lights, appears musty and blue-tinged at first, but later dark brown, with the ridged qualities of a muddy field, ploughed up and dried out. Below each tree is a mound (another doubling) made of the same material as the stage cloth.

The acting style of *Mushroom Language* is melodramatic. The powerful performer energy is excessive, surpassing the function of the action. It is the energy of the popular performer, the entertainer; the actors are dressed in bland white shorts and shirts, often covering their faces with stockings, masks, or probosces. As the action develops, their familiar "variety" hyperenergy is – like Beckett's *Godot* characters – expressed in episodes which each time fail to sustain a human performance.

While the performers at times enact the hollow stereotypes of a bickering couple or a parent with a demanding child – "again, again!" – much of their performance energy is expended in imitating various kinds of fungi or fungal hybrids like lichen. Whatever is there of human nature is in decay, crumbling, and decaying. Subtitled "*Gothic*," the performance has the animalistic de-psychologized, and desire-driven quality of late-romanticist drama but pushed beyond Gothic's now comfortable and commodified morbidity and deeper into the loam. What *Mushroom Language* offers is an option to perform a dismantling of humanist performance, an undoing that could formerly serve Samuel Beckett's absurdist vision, but – in its disintegration and decay – can now help a diagrammatical performer meld the detritus of traditional and modernist acting to a collaboration with more-than-human others.

In *Mushroom Language*, the performers pull materials that double for mycorrhizal webs or hyphae from openings in the floorcloth, yet repeatedly some remnant of human identity sucks them back to enacting the tortuous remains of a romantic human attachment. All the time psychological decay gnaws at their identities; their performances rot; they shuffle through a variety of dramaturgical ploys – storytelling, portentous folkloric masking, comedic obscenity, the flesh-tattered bones of domestic naturalism, Gothic horror – but these are all in decay, incapable of meshing the human with the more-than-human or transcending from one to the other.

Rather than languishing in *jouissance* – by owning the pleasure of stinking decay or capitalizing on their virtuoso mimicking of lichen, truffles and the exploding spores of *phallus impudicis* – Ali Matthews and co-performer Tom Halls end diagrammatically by pulling their whole imperfect effort through 90 degrees, turning the entire stage picture diagrammatical. They pick up the front edges of the floorcloth and raise them up high so the performers disappear and then drag the floorcloth back beneath the branches of the *Godot* trees, sweeping up discarded spore-balls, mud-coloured mounds, and masks of pig and crow. For a few seconds the stage cloth bucks and writhes like vegetal monsters as the sound system plays the repeated chorus lyric "mulch me, go on without me." A seething loamy liveliness, this is also a vision of an abyss, of life born "astride a grave" (Beckett, 1954, p. 57).

More even than a "worlding," this metaphorical death of the performance is a vision of limitless creaturely connections and of an absurdist irreversible beyond-life. A queering of post-humanity is abruptly realized. All the previous failed approximations in the performance of *Mushroom Language* are overcome by a scenographic action in which the mess of spores and masks becomes a diagram, turned on its edge. The comedic and incongruous energy of the human performers – rehearsing their deaths and their beyond deaths, learning to eat through their skin so they become (as in Samuel Beckett's *Not I*) only mouth – passes beyond virtuosity and failure into the fecund multiplicity-anonymity that the "characters" have aspired to and repeatedly failed to be. They succeed by failing and they disappear into the diagram. In that moment a conventional theatre studio, briefly, becomes a "three-dimensional metaphor."

(After reading these two How To Act sections, check back to the Camouflage and Transparency section in Chapter 3.)

Narratives

You have no need of an overarching narrative; it will likely pull you towards a character drama. Instead, look for small and contained stories that you can use in episodic chunks. Consider the narratives as you might any other objects in the performance, as somewhat withdrawn, as having their own integrity and discreteness. Rather than a uniformly woven texture (which is what a playwright is often attempting), it is better to think of the performance as an orrery – a circulation of things each with their own gravitational pulls – that demonstrates the relationships between all the parts, rather than too fine a weave.

Repetitions

In diagrammatical performances, repetition and difference are the parts of a dramaturgical dynamic that replace the "what is at stake" of character drama. In his work on Brecht's Epic theatre - with its dual priorities of presentation of social reality and the processes of change within performance itself – Walter Benjamin turned to children's play (something that is crucial to our remaking of ritual for diagrammatical performance, see The wandering voice section above) for insights into this dynamic.

Repetitions flatten time, as they "maintain[s] a simultaneous reference to the past and to the future" (Steinhauser, 2016, p. 18), and this invites a kind of change that is not tethered to the forward rush of linearity. To describe this kind of change, Walter Benjamin referred to his observation of children's play; how in a game's many repetitions, what he calls a child's "genius of variation" was brought to bear without disrupting the repetitions of the playing itself. He noted how "children combine with thousands of surprising variations the taught gestures into a theatrical totality" (Steinhauser, 2016, p. 26).

This form of serious but play-like devising is what a diagrammatical performance maker should attempt. To choose elements and then bring them into the devising space and repeat them, noting whatever "variations" step forward during the game of playing with these matters. Not trying to shape a performance from overhead to fit a totality but repeating your materials and staying attentive to what they produce. Deferring your consideration of the "theatrical totality" and letting the materials be the *genii* of variations; allowing a totality to emerge. Attention to the more-than-human – even in unexpected forms such as the viscosity of treacle or the changing form of a gesture – is crucial here; this attention includes taking things seriously, so that when variations appear – subtle, absurd, clunky, barely visible – they are taken on board, noted, and repeated within the emerging "totality," driving the move towards a "totality."

Coincidences

Take notice of coincidences. They may be the "meshwork" talking to you in a language of events and connections. While the contents of these experiences may be important so also may be the pattern of coincidences; you may be being given a pattern for your performance.

Allies

Develop an eye for allies. Those things that have a thickness to them, that wear their history; maybe ask them to be a part of your performances? Deploy empathy. Ally with the unwanted, the discarded, the things washed up on beaches, found in gutters, stuff from spills and slag heaps; there's a good chance that what capitalism does not value is gold to you. Fallen signposts, eroded lettering, paint scraped into abstractions; visit charity shops and scrapstores.

Assemble your allies over the years. Live beside them, work with them variously – in installations, exhibitions, performances-without-audiences (let's call them "rituals") – listen to stories that gravitate to them, attend to impressions and associations that arise from them, get used to a feel for them.

Be painterly

Work your materials. Code your colours. If you ever arrive at representation, only do so by accident or in despair of all other methods. Make fields in the cosmological (not the agricultural) sense. Be painterly in process; you are not trying to be decorative.

Text

It is probably inappropriate to say "script." Unlike a play, a diagrammatical performance is not driven by a script. Any text will be made up of any number of different components. In our experience, much of any text that goes into devising will be scrapped before performance; mostly replaced by being better expressed by an action or a thing. In this case "better" means more connectively. Unlike text-based theatre, where the words are often the determining entity, and a vehicle for the performances' longevity, there are no such relations within diagrammatical performances. Instead, the criteria at work are those that best connect a performance to its more-than-human allies. We do not have a hard and fast prescription for how to arrive at that. In other words, you will need to experiment, heuristically, to find out what works best for you.

There are, however, things you can try, though they carry no guarantee with them. These include:

Automatic writing: take yourself to a place that you want to perform with and just write, as unselfconsciously as you can. Allow the place to write you. See what comes.

Quotation: gather texts from your places. Or from around the allies. These might be from information boards, overheard conversations, stories in a café, local history pamphlets, anything...

Cut up: once you have gathered texts, devise some system for assembling them, or cutting them up and juxtaposing their parts with each other. Does some secondary sense emerge?

Ashes: after writing (unexpectedly) a large number of poems about places we were visiting, we gathered our poems together and cut them into short phrases. We reassembled some of these into new poems. We retained many of the short phrases and used them as gifts, inviting audiences or visitors to choose a phrase at random from a hat full of the tiny slips of paper. Later we burned a large quantity of these slips, and used their ashes in performances, as material to make shapes with stencils. Through fire and burning we were able to transform not just the genre or form, but also the material content of the art.

Diagrammatical gestures

You might choose a daily practice, in common with the kinds of disciplines that Grotowski's Laboratorium and Odin Teatret drew upon, or such as those of traditional theatres like Kathakali. However, like us, you may eschew performance-training in favour of a place-based sensory practice, going daily to "be there" with others. Nevertheless, a diagrammatical performance is a performance, and that means that all the actions you perform will be recognized within a web of meanings, in the same way as any words will be recognized as part of a language of signifiers.

In *Scoriton Mysteries* we took two gestures – "found gestures" – from documentary footage from 1965 of the rituals of the Aetherius Society (a flying saucer religion worshipping alien "space brothers" based primarily on the theosophist belief in ascended masters of its founder, George King). These were gestures of connection; one invoking aliens to come from above, the second directing "prayer energy" into a "prayer battery," where the energy was stored for use later (in 1965 to prevent nuclear weapons testing and to ward off a pandemic "from the East"). While we did not parody the Aetherius ritual, and described their aspirations in a straightforward manner, we were enacting a kind of "empty magic" in that the gesture was not only "loaded" with meaning but without any expectation of an effect. Instead, these were "allowing" gestures, inviting the agency of "others" without requiring them to act, arrive, redeem, or identify themselves in any precise way.

Interestingly, the entire audience of around a hundred people raised their hands in the gestures. Somewhat like the operation of "thickness" in objects, there was something in the affordance of these gestures, sufficiently emptied, for the audience to step into them. Later the audience would similarly eat wafer-like "Flying Saucer" sweets (was this an alien communion?) and communally hold a green ribbon to make a "pixie ring" (a notional representation of the circle of altered grass resulting from fungal growth).

The quality of these gestures is that they are almost but not quite empty ones. In *A Séance For Stonehouse* there was no striptease; sex work was invoked by Helen hobbling round the space, assisted by Phil, in unfeasibly high heels. Phil placed a white muslin over Helen's face, Helen seated in a chair within the séance/audience circle of chairs, and then handed her a linen homunculus (made by Phil's daughter, artist Rachel Sved) to cradle in her arms. Part icon, part sex worker; the image drew on the skeleton-baby in Odin Teatret's *Il Millione* (1978), evocative of the children born after a Carnival nine-months previous. The gestures – the tottering, the making of the mini-tableau – are recognizable, but there is very little to them; they are designedly emptied in order for the audience to fill or thicken them with a multitude of their own associations. They are sufficiently shaped, they have an outline, and certain lines and two-dimensional markers – like diagrams – to hold those associations meaningfully, but not restrictively.

So the diagrammatical gesture is one that has been hollowed out, diluted, or blanched sufficiently to allow an audience's associations to assume a degree of autonomy. It adopts the notion of the agency of ideas from Jamie Mcphie (2018) or of memes (as exampled with the flying saucer shape in the *Scoriton Mysteries* case study in **Chapter 6** above). The diagrammatical gesture thus invites associations and notions into the web of performance, as autonomous agents. The diagrammatical gesture does not invite a precise association, it is not designed to be recognizable in a particular way, it does not add to a set of (variably or precisely) predictable responses determining to some extent the reception of the performance of it. It is possible that an audience may experience common associations (after all, our lexicons and codes are in the first instance drawn from the world), but that is not the intention or function of a dilated diagrammatical gesture.

A diagrammatical performer, may take something from the stagehand-like behaviour of a performer in a non-matrixed "happening" (Kirby, 1972/1995), and something from Brecht's Epic Theatre, for which, Walter Benjamin argues, the performer should "block out" their gestures as a twentieth century typesetter would assemble or "block" out letters and words for a printing press, putting them together in units, with an eye to their local sense rather than their charisma or flow; prioritizing "a 'chunky' continuity" (Steinhauser, 2016, p. 21).

Rather than carrying a message, or representing an emotion or point of view, the diagrammatical gesture is an invitation, a turning inside out to welcome any response (from examination to surgery). After giving the instructions for the "invoking aliens" gesture to the audience, Helen said "we'll see what happens" and the audience laughed, perhaps to acknowledge that none of us were expecting anything to "happen." The "empty gesture" thus opens up a space of welcoming ambiguity (rather than any particular meaning or radical destabilization of meanings) into which an audience is free to associate.

Given that diagrammatical gestures are defined by what they are connective to in the moment, rather than any technical prowess or representational virtuosity, there are no exercises to practice in order to get better with them. The gestures may be found by attention to existing or "found" gestures; whether that be from UFO cults or car maintenance or gymnastics or whatever. Their crucial quality is not their level of physical difficulty, representational veracity or elegance. A certain clumsiness might be useful in warding off any focus on technical efficiency. Finding the right gesture will mostly come down to one's judgement; intuiting its degree of invitation and the space for hosting within it.

And finally...

Over time, things get easier. If you are willing to experience periodic upheavals, to shift the usual and then stabilize; to deterritorialize and then reterritorialize. These are necessary processes for staying sensitive. It is easy to give too much advice about being intuitive; in the end, everyone has to find their own way and develop their own senses. This cannot be a toolkit for performances that are anything like Crab & Bee's. Even your repertoire you will have to keep re-inventing, re-calibrating, because most of it is not human. Most of it has yet to be performed. Excited and excitable times are coming.

References

Allen, R. (2016) 'The theatricality of objects: Object theatre beyond the puppet'. *Ånd I Hanske*. 34(1), pp. 6–13.

Augé, M. (1995) *Non-places: Introduction to an anthropology of supermodernity*. London: Verso.

Beckett, S. (1954) *Waiting for Godot*. New York: Grove Press.

Bennett, J. (2010) *Vibrant matter: A political ecology of things*. Durham and London: Duke University Press.

Crab & Bee (Helen Billinghurst & Phil Smith). (2020) *The pattern*. Axminster: Triarchy Press.

DeLanda, M. (2013) *Intensive science and virtual philosophy*. London: Bloomsbury Academic.

Deleuze, G. and Guattari, F. (1988) *A thousand plateaus*. Translated by Brian Massumi. London: Continuum.

Kaprow, A. (2003) *Essays on the blurring of art and life*. Berkeley & Los Angeles: University of California Press.

Kershaw, B. (2007) *Theatre ecology*. Cambridge: Cambridge University Press.

Kirby, M. 1972/1995. 'On acting and not acting', in Phillip Zarrilli (ed.), *Acting (re) considered: A theoretical and practical guide*. London & New York: Routledge, pp. 40–52.

Latour, B. (2004) *Politics of nature: How to bring the sciences into democracy*. Cambridge (MS): Harvard University Press.

Mirčev, A. (2024) *The poetics of performance diagrams*. Cambridge: Cambridge University Press.

O'Sullivan, S. (2024) *On theory-fiction and other genres*. London: Palgrave Macmillan.

Schechner, R. (1993) *The future of ritual*. London & New York: Routledge.

Simondon, G. 2011. *On the mode of existence of technical objects* [extract]. Translated by Ninian Mellamphy. Edinburgh: Edinburgh University Press.

Smith, P. (2025) 'Three-dimensional metaphor: After site', in Victoria Hunter & Cathy Turner (eds.) *The Routledge companion to site-specific performance*. New York & London: Routledge, pp. 56–67.

Smithson, R. (1996) 'A provisional theory of non-sites', in Jack Flam (ed.) *Robert Smithson: The collected writings*. Berkeley & Los Angeles: University of California Press.

Steinhauser, S. (2016) 'Of secret signals, absent masters and the trembling of the contours', in Eirini Kartsaki (ed.) *On repetition*. Bristol: Intellect, pp. 15–32.

Tsing, A.L. (2015) *The mushroom at the end of the world*. Princeton & Oxford: Princeton University Press.

8 Conclusion

In 2007, Baz Kershaw published his book *Theatre Ecology*. Rather than assuming that a progressive theatre could help in a response to a threatening climate change, he characterized theatre as inimical to any close relationship with the ecological meshwork and as a part of the problem, not of a solution. He suggested that, having reduced nature to a representation, humans present their environment as though they are no longer in it; so that words like "nature" and, even, nature – both representation and actuality – have become constructs of a separation in which theatre is conceptually and materially stuck thanks to its most valued qualities of imitation and representation.

Kershaw extended his critique of theatre to its anthropocentrism, ascribing "the primary productive purpose of theatre.... *not* as the staging of performances that represent 'landscape' or 'wilderness' or other constructs of 'nature'.... but to the creation of spectatorship" (2007, p. 305). He then charged theatre with being "the social institution that has most quintessentially modelled the abstraction of humans in the 'natural world'.... the modernist culture-nature binary.... has been the principal misconception causing humans to destroy the Earth's environment" (p. 306).

This characterization of theatre put Kershaw in a difficult position. For his wider description of society (in tune with arguments we have shared in Chapter 3, section "The cancellation of the future") is that "by the third millennium new global forms of political, economic, mediatized and technological change together engendered the 'performative societies': societies that are crucially constituted by performance" (2007, pp. 11–12). Given that contextualization, it was hard to see on what surplus resources theatre performance could now draw to remove itself from the vicious recycling of its ideological role. (Andrej Mirčev comes close to constructing the same conundrum for diagrammatics, when in *The Poetics Of Performance Diagrams* [2024] he associates "a world increasingly regulated and governed by algorithmic protocols" with "a *diagrammatic effect* on the real" [2024, p. 69]).

Kershaw seemed to acknowledge the paradox when he wrote of being "sceptical about theatre's purposes, even to being 'anti-theatre'" (2007, p. 306). By spreading his critique to a "performative society," even a sidestepping of theatre into other non-theatrical or post-dramatic performance

DOI: 10.4324/9781003606673-9

forms – such as a live art, for example – seems to fall foul of the same problem of spectatorial production that Kershaw (and others) put at the heart of theatre's inadequacy.

This – and the social and cultural context that Kershaw intuits – puts us, along with other performers, into a challenging dynamic; a postmodern circularity. Even when eco-critical theorists advance a hopeful path, this cyclical shape threatens to return. So when Una Chaudhuri claimed that "[B]y making a space on its stage for ongoing acknowledgements of the rupture it participates in – the rupture between nature and culture, forest and books.... the theatre can become the site of a much needed ecological consciousness" (1994, p. 28), the resulting consciousness, problematically, is still human and spectatorial. Despite the virtue of its eco-thinking, this repeats the rupture.

To avoid a shifting of the problem from a planet burning up to a failure of critical thinking (which is a much more comfortable ground for critical theorists), or to "constructs" that forefront unintended parodies of nature in anthropocentric discourse, it is the diagram that has, for us, proved a useful "carrier bag" for gathering unruly resources for a resistance by matter. A diagram's abstract qualities map those constructs of "nature", the production of which Baz Kershaw describes. While the diagram's creaturely qualities facilitate a libidinal "naturally-unnatural" energy capable of escaping modernist and postmodernist grids onto the more hospitable grounds of hybrid diagrams described by Andrej Mirčev as a "multisensorial scripturality of graphic traces and affective embodiment" (Mirčev, 2024, p. 1).

Mirčev – despite his pessimistic comments about "a diagrammatic effect" (above) – resuscitates the diagram wherever there is "breaking through the dualism separating words from images and subjects from objects". This is when "diagrams and diagrammatic imagination allow for envisioning new scenarios" (p. 9). If such a shift to an integrated physicality/materiality, from fetishized literary and critical grounds, sounds something like Antonin Artaud's demand in *The Theatre And Its Double* that "[W]e must admit theatre's sphere is physical and plastic, not psychological" (1970, p. 52), then it is such only in so far as it stands back from Artaud's pursuit of an ecstatic and transcendental anthropocentric meta-drama. In contrast, we assert our loyalty to a "dull matter (it, things)" – that we do not distinguish from "vibrant life (us, beings)" (Bennett, 2010, p. vii) – over an idealism, whether it be revolutionary or mystical. We favour above all else matter's slow mutations; a move from an art practice to an art of living. In this we are following Tadeusz Kantor's suggestion that "in a 'dense' life situation a work of art ceases to have its own significance" necessitating a shift to "where the art object could be replaced by a real-life object" (Miklaszewski, 2022, p. 82); a positive take on Kershaw's "anti-theatre."

Yet alternative Western theatre has tended to be more comfortable with Artaud's ecstatic vision than with an address to a physical environment. In 2014, Wallace Heim expressed some surprise and disappointment at how "vexed" were attempts through the Ashden Directory of Environment and

Performance (edited by Heim and theatre critic Robert Butler) to encourage environmentally aware productions. Heim speculated that perhaps "the environment or ecology was too materially "real" for theatre, confounding the usual, more comfortable combinations of reality and fiction" (2014, p. 6).

Yet matter matters. In *Theatre Ecology*, Baz Kershaw attends closely to institutional and commercial spaces of constructed "nature" such as zoos, the Eden Project in Cornwall (UK) and Biosphere II (Arizona, USA); "a giant glass 'ark' the size of a huge aircraft hangar" in the desert (2007, p. 300). Our chosen grounds, however, have been the "edgelands" and "voids" on the peripheries of cities, ghost villages abandoned due to mining, and verges unintentionally rewilded thanks to municipal neglect. We take our allies from the spacecraft-shaped trash we find in street gutters, or from the bones we gather off the hills. We have tried to avoid the problem for eco-theatre articulated by Wendy Aarons and Theresa J. May as how "[E]cological stories take place on a scale beyond the human" (2012, p. 4), by allying with things close to hand, aware of them as parts of hyperobjects that connect us to the bigger scale but without the temptation to Artaudian self-obliteration or transcendence. In our aesthetics, "cosmic" is something that begins close to hand, rather than in a mind blown.

On one of our diagrammatical walks, for our local Village Hub community organization, we pointed to the forget-me-nots growing in the gaps between the granite paving stones. We asked participants: "what is that? A brief flowering that will soon disappear from the street? Or a meadow from which the street will soon be gone?"

In an article that deploys and adapts Gregory Bateson's diagram of "The Dynamics Of Ecological Crisis" (2000), Baz Kershaw writes in "Projecting Climate Scenarios, Landscaping Nature, And Knowing Performance: On Becoming Performed By Ecology" (2018) of a way out of the recycling of rupture, by transforming a commonsense that "ecological processes... are obviously crucial to life on the planet" into a "knowing performance... [T]inged with a hunch that we are somehow performed by ecologies" (2018, p. 43).

In June 2010, while cutting a meadow on his smallholding in Devon (UK), by scythe – avoiding the "massacre of micro-creatures" caused by motor-mowing – Kershaw, lost in memories of "trance-like dancing," left a series of winding paths rather than straight lines. On discovering an equivalent to his mown meanders in global flows, this "accident" initiated his *Earthrise Repair Shop* project. This began in earnest with his asking "twenty artists, scholars, friends and family" to "try out" the meander in the meadow; leaving them free to dance, walk, explore, compose, observe, lie down and sleep, draw, sit, think and respond to the meadow as they wished. Which led to further meadow meanders – "like most traditional mazes and labyrinths.... prov[ing] accessible across especially diverse individuals, generations and cultures" – being cut in St George's Field in Leeds, UK (2012), at Tempelhof Feld eco-park in Berlin, twice at Warwick University (2014 and 2015) and in

Saltaire, UK (2015), with some 1200 people participating (Kershaw, 2018, pp. 52–53).

Only while typing the first draft of the above paragraph did Phil remember that he had been one of the "twenty." He found on his computer a file entitled "visit to Baz"; his response posted to Kershaw back in 2011. Among its reflections are these:

> 'The maze pushes the focus outwards. It allows little introspection', '[T]he maze slopes and constantly challenges the body to shift its centre of gravity in response. The maze is somatic.... The maze "empties" the mind – by shaking the brain it opens up coagulated gaps, rendering them vulnerable to an outside that rushes in to fill the tiny sucking vacuums'.

A sample size of one is hardly reliable… yet Baz Kershaw's eschewing of institutionalized constructs of "nature," integrating dramaturgy and meadow, diagramming global flows, and then setting visitors/performers/spectators loose within a "drama" of grasses and flowers, under a cyclorama of sky and beside a proscenium of trees, had had its effect on this poorly representative sample: "I am not very affected by this," Phil wrote, ironically, "except for this prosthesis that I have now got attached to me. Thrashing about" (Smith, 2011, private correspondence, unpaginated). The writer of that note in 2011, at the time deeply sceptical of the effectiveness of eco-theatre, is now co-authoring this book.

Where Crab & Bee might depart from the arguments in Baz Kershaw's 2018 article is his adding the idea of "ecoumility" (as a means to "ecosanity") to the "hubris" on Bateson's diagram. For it suggests to us too much a return to a normative "balance" or "green husbandry." In the article itself, this leads Kershaw to defend "ecoumility," as a means to "ecosanity," by retreating from his "corporeally embedded scenarios" (like his meadow meanders) to a strategy for "how literature and performance arts equally might evoke quotidian natural processes re-experienced through the prisms of human creativity" (2018, p. 58).

Instead of such "prisms," however, we have chosen to go another route. We have adapted a creaturely diagram via the schizoanalysis of Deleuze and Guattari to pursue a perverse-in-nature, a cosmos as unevenness, eschewing "filiative production or hereditary reproduction" in favour of the "naturally unnatural" where "there are as many sexes as there are terms in symbiosis, as many differences as elements contributing to a contagion" (1988, p. 267).

While it might be frustrating for theatre and performance makers, the idea of a dispersal – sideways – beyond the architectural and conceptual boundaries of a performance space is not intended to add extra cultural capital to the art of performance. That is part of the economy of site-based performance, at its most concentrated in an "adaptive" form – a means of adapting a performance from one unconventional site to tour to many, transferring site-capital in one place into performance value at others – in which Melanie Kloetzel detected a dominant "neoliberal drive towards ecological and economic precarity.... that tends toward the spectacular" (2017, pp. 111–112). While a diagrammatical performance often displaces its origins in a located "being there" with a place or places, it seeks to leave most of, if not all, the value with the original places and their material agencies.

In 2019, as part of a week-long performed Crab & Bee residency on the ruined Second World War concrete slipway at Teats Hill in Plymouth, we sought to tell the story of the arrival from Syria of Albina and Her Sisters. This origin story of Albion/England, Albina being the mother of the local legendary folk monster Gogmagog, was introduced to the diverse working class community there. Often with local families, we made an icon of Albina, using various materials from the beach there – rocks, broken glass, seaweed – and painted tentacles on the slipway in a white China clay mixture. We wrapped huge blocks of granite and limestone in large white and red cloths; we lit fires, carried seawater, and ran it down the slipway. We built and maintained a shrine in a small cave, adorning it with white pebbles, paper boats, candles, and milk (the milk of Albina of Teats Hill) and scattering it with China clay powder. We used the red and white cloths as enormous flags.

Throughout this time, we were a visible presence to residents of the nearby housing association flats. However, rather than attempting (in the way of the missionary) to insist on the participation of local people, we took our time and carried out our business of just "doing stuff with stuff". Eventually, curiosity took its course, and residents of Teats Hill came down to the slipway to find out what we were doing. Then they would join in with our makings, add offerings into the shrine, share in our ritual fires and games. We would share our story of Albina, and very often they would share a story with us in return. On our last day, just as we were about to dismantle our assembled makings and pack up, a mother breastfed her baby on the slipway, a white milky sea mist rolled in, and a fleet of white swans swam up to the end of the slipway.

As with the leaves at Tincombe, or the cylinder of dust and hair rolling across the floor of the room in the old pub during *A Séance For Stonehouse*, the capital of performance was lost to us, Crab & Bee, and was taken up by the more-than-human agencies of the place. Given the way these things happen coincidentally, unrepeatably and by their own volition, they can sound like descriptions of something "magical" or "supernatural," but that may be because globalized humans tend not to recognize the agency of things they would otherwise characterize as "inert," "inanimate," and lacking cultural value until deployed by human artists. Within a diagrammatical performance practice, such moments, while exceptional, are what we aspire to make the rule. To open space for the more-than-human performers to keep or generate their own value. So an eco-diagrammatical performance is a performance with and by ecological agents, not one that is about or representational of them.

In such performance, in which there is a draining or displacement of familiar human cultural values, there is an opportunity for human performers to experience a hybrid "becoming there" – in which the werewolf ("the call to become wolf" [Deleuze and Guattari, 1988, p. 32]) and the centaur ("monster more intelligent than the most intelligent of men, the image of the marvellous dissimulation of signs into one another" [Irigaray, 1993, p. 81]) become possible. In each case, there is displacement from the spotlit performer to the web or collective animal. In the case of the wolf "you can't be one wolf, you're always eight or nine... the important thing is the position of the mass.... the subject in relation to the pack or wolf-multiplicity" (Deleuze and Guattari, 1988, p. 32). In the case of the centaur, it is a "monster with which we make a libidinal pact, neither destroying us nor we him, not to take him as lover but ourselves as becoming libidinally-larval" (MacCormack, 2008, p. 85).

The surplus of any such effective eco-diagrammatical performance is not a cultural one, but – in part – a schizo-psychological gain, in which the transformative profit of performer and audience, through their "becoming other" is offset by the transfer of at least half to the others, in the form of a pack/multiplicity and a process "at the point of de-humanization" (MacCormack, 2008, p. 85).

This has led us to a marginal terrain, somewhere between the meadow and the constructs of "nature." We go there in search of "a secret signal of the to-come" that is accessible through the gestures of gaming and playing, as identified by Walter Benjamin in his study of children's play (2005). Such a diagrammatized gesture always being in "excess of its relative determinability as a sign," while through its "inaccessibility and withdrawal from or supersession of present sense and tense, the secret of the to-come nevertheless signals itself." As with the withdrawn object of Object Orientated Ontology, it is thanks to this withdrawal from availability that "the child is able to gesture beyond itself" (Steinhauser, 2016, p. 26).

Amplified in marginal spaces where modernist ruins and post-industrial sites are spontaneously rewilding, the "withdrawal" of such objects is similar to the child's gesture hosting a secret signal. Their ontological withdrawal is amplified by their ejection from processes of production, distribution, and consumption. Like the child's play gestures, these discarded objects, via their inaccessibility, are a haunting of the present by the future. Given the mechanics of decay and industrial decline, abject spaces have become laboratories of mutation, emitting signals from the future to be decoded as findings.

It is thus that we can retain optimistic – or at least equitable – information, working on the basis that there is no tech fix, no likely global superpowers' reparative action, nor any magical-theological amelioration coming. Instead, we can work (making artworks with material allies) on the basis of a Planet B to-come. Ours is a non-mechanistic and leaky futurism in which knowing how to get on with strange insects and work around other mammals is likely to be more useful than data management skills.

We may be wrong. If we are not wrong, then libidinal energy (desire) will continue to be crucial to living, anyway. Against transcendental spiritualism and mechanistic transhumanism, we have taken gravity as an ally and a philosophy; the attraction of all things to all other things and of our bodies to the Earth. This is not an anti-scientific vision. If we (or, rather, those far more experimentally adept than us) can grasp the composition of gravity waves, and learn how to use them as communication, then our connections may be even more extensive, multiplicitous, and queer than we have suggested so far… who knows… As for now, we prefer to sink with what matters and explore death-in-life by embracing composting and de/composing.

Our "diagrammatical performances" have emerged as a result of many meetings: with trees whose roots lift up the pavements, with an exhausted fallow stag recovering in the shadow of decaying garages (guarded by a human ally from a nearby house), with returning meadows on verges and in housing estates with hidden woodlands, with herds of wild ponies grazing around Dartmoor's tors. On the one hand, we have been responding to human degradations of an environment – the pollarding of trees during nesting season, the relentless building over meadows, encounters with long ago imported giant Pacific Oysters on the beach opposite the docks with their rotting nuclear submarines (reactors just about intact) – but even more so to the billions of worms turbating in the mud of long abandoned industrial creeks, and to half-imagined personalities of place seen out of a corner of the eye and returning in our dreams.

Rather than a normalization or naturalization, in response to the "artificialities" of our urban lives, we have more often found ourselves reacting to weird or *unheimlich* more-than-human actors that we encounter in the "natural" environment. Like the clumps of forget-me-nots that became *unheimlich* actors on our diagrammatical performance walk in 2024, threatening the homes of our audience members, burying our futures under their coming meadow. Likewise, the fallow deer clattering along the corridors inside the local hospital, or the subsiding red sandstone cliffs covered with "garden escapees" that threaten all the time to engulf the sole coastal railway line linking us to the rest of England, or the giant herring gulls driven inland by the fishing out of the herring and now battling locals and tourists for their ice creams and bags of chips. These are all daunting players in our eco-drama; heralds of far greater disruptions yet to come. Thus, a diagrammatical performance practice, as we describe it here, will always be primarily shaped by particular environments and in some form of alliance (or "alience") with our more-than-human others.

The defining elements of this performance making will always be first an attending to more-than-human places and entities and then – sidestepping the site-specific or immersive urge to make them our performance spaces – a resorting to "non-sites" or "three-dimensional metaphors"; inviting more-than-human agents (from "inanimate" minerals, skins, feathers and bones to icons of wolf-saints, invoked packs and *genii loci*) to be our leading characters. The diagrammatical performances we have described and discussed here eschew any pursuit of efficacy or high aesthetical value; they are something closer to the studied indifference to dramatic arts of companies like Index Theatre and Forced Entertainment: "unwilling or unable to work with theatre's codes to satisfy that medium's problematically social audience" (Etchells, 1995, unpaginated).

It is, however, to Baz Kershaw's "meadow meander" that we maybe come closest. Without access to a meadow, you could draw a similar diagram in your back yard or on a grass verge. Kershaw's diagram performs with a large more-than-human ally (grass) and its human arranger quickly absents themselves to invite others to perform and discover a relation to the primary performer, the meadow. The dramaturgy is largely one of grass and how it lies; the nature of the diagram Kershaw kept largely to himself (it was a model of the world's oceanic flows) yet it was always there to be seen, danced and touched: a three-dimensional metaphor of oceans in the "non-site" of a meadow. Wind, rain, odour of trees, sounds from branches, tractors on the nearby road (an old drovers' route) all played their parts, but the meadow-as-ocean-brain dominated and enfolded everything. As good a model as any for how to involve a performance in its materials, and at the same time align its metaphor diagrammatically with its materials, so they sink into each other.

In conclusion, however, like a good diagrammatical performance, we end not with a virtuosic flourish, or a sudden acceleration to optimism, but with a salute to the ordinary/extraordinary materials – objects and entities – that we and others have been joined with in performances: grass, branches, pixie rings, white clay powder, red earth, crows, sun-bleached bones, gold, black treacle, flying saucer-shaped trash, tin, lead, sheepskin, rust, the ash of burned poems, dustbin lids, a fleet of swans, red high-heeled shoes, a fabric homunculus, green and red ribbons, paper boats, traffic cone, painted shrine, alchemical scaling of metals, chalk, threads…

References

Aarons, W. and May, T.J. (2012) *Readings in performance and ecology*. Basingstoke: Palgrave Macmillan.

Artaud, A. 1970. *The theatre and its double*. Translated by Victor Corti. Montreal, London & New York: Calder.

Bateson, G. (2000) *Steps to an ecology of mind*. Chicago: University of Chicago Press.

Benjamin, W. 2005. 'Program for a proletarian Children's Theatre', in Michael W. Jennings, Howard Eland and Gary Smith (eds.), *Walter Benjamin: Selected writings, volume 2: Part 1, 1927–1930*. Cambridge: Harvard University Press.

Bennett, J. (2010) *Vibrant matter*. Durham & London: Duke University Press.

Chaudhuri, U. (1994) 'There must be a lot of fish in that Lake": Towards', *Theatre*, 25(1), pp. 23–31.

Deleuze, G. and Guattari, F. (1988) *A thousand plateaus*. Translated by Brian Massumi. London: Continuum.

Etchells, T. (1995) 'Index Theatre'. *Frieze Magazine*. Posted 6.1.95. Available online at: https://www.frieze.com/article/index-theatre. Accessed on: 23.2.25.

Heim, W. (2014) 'The continuing lure of ecological performance', in Wallace Heim and Eleanor Margolies (eds.) *Landing stages. Selections from the Ashden directory of environment and performance, 2000-2014*. London: Crinkle Crankle Press.

Irigaray, L. (1993) *An ethics of sexual difference*. Translated by Carolyn Burke and Gillian Gill. Ithaca: Cornell University Press.

Kershaw, B. (2007) *Theatre ecology*. Cambridge: Cambridge University Press.

Kershaw, B. (2018) 'Projecting climate scenarios, landscaping nature, and knowing performance: On becoming performed by ecology', in Carl Lavery (ed.) *Performance and ecology: What can theatre do?* London & New York: Routledge.

Kloetzel, M. (2017) 'Site, adapt, perform: A practice-as-research confrontation with climate Change', *Dance Research*, 35(1), pp. 111–129.

MacCormack, P. (2008) *Cinesexuality*. London & New York: Routledge.

Miklaszewski, K. (2022) *Encounters with Tadeusz Kantor*. London & New York: Routledge.

Mirčev, A. (2024) *The poetics of performance diagrams*. Cambridge: Cambridge University Press.

Steinhauser, S. (2016) 'Of secret signals, absent masters and the trembling of the contours', in Eirini Kartsaki (ed.) *On repetition*. Bristol: Intellect, pp. 15–32.

Index

For Product Safety Concerns and Information please contact our EU representative GPSR@taylorandfrancis.com Taylor & Francis Verlag GmbH, Kaufingerstraße 24, 80331 München, Germany